T0192046

UNITEXT for Physics

Series editors

Michele Cini, Roma, Italy
Attilio Ferrari, Torino, Italy
Stefano Forte, Milano, Italy
G. Montagna, Pavia, Italy
Oreste Nicrosini, Pavia, Italy
Luca Peliti, Napoli, Italy
Alberto Rotondi, Pavia, Italy

More information about this series at http://www.springer.com/series/13351

Carlo Alabiso · Ittay Weiss

A Primer on Hilbert Space Theory

Linear Spaces, Topological Spaces, Metric Spaces, Normed Spaces, and Topological Groups

Carlo Alabiso
Department of Physics
University of Parma
Parma
Italy

Ittay Weiss
South Pacific University
Suva
Fiji

ISSN 2198-7882
ISSN 2198-7890 (electronic)
ISBN 978-3-319-35351-7
ISBN 978-3-319-03713-4 (eBook)
DOI 10.1007/978-3-319-03713-4

Springer Cham Heidelberg New York Dordrecht London

© Springer International Publishing Switzerland 2015
Softcover reprint of the hardcover 1st edition 2015
This work is subject to copyright. All rights are reserved by the Publisher, whether the whole or part of the material is concerned, specifically the rights of translation, reprinting, reuse of illustrations, recitation, broadcasting, reproduction on microfilms or in any other physical way, and transmission or information storage and retrieval, electronic adaptation, computer software, or by similar or dissimilar methodology now known or hereafter developed. Exempted from this legal reservation are brief excerpts in connection with reviews or scholarly analysis or material supplied specifically for the purpose of being entered and executed on a computer system, for exclusive use by the purchaser of the work. Duplication of this publication or parts thereof is permitted only under the provisions of the Copyright Law of the Publisher's location, in its current version, and permission for use must always be obtained from Springer. Permissions for use may be obtained through RightsLink at the Copyright Clearance Center. Violations are liable to prosecution under the respective Copyright Law.
The use of general descriptive names, registered names, trademarks, service marks, etc. in this publication does not imply, even in the absence of a specific statement, that such names are exempt from the relevant protective laws and regulations and therefore free for general use.
While the advice and information in this book are believed to be true and accurate at the date of publication, neither the authors nor the editors nor the publisher can accept any legal responsibility for any errors or omissions that may be made. The publisher makes no warranty, express or implied, with respect to the material contained herein.

Printed on acid-free paper

Springer is part of Springer Science+Business Media (www.springer.com)

Ittay Weiss dedicates the book to his daughters, Eugenia and Esther, for allowing him to work on it.

Preface

About the Book

This textbook is a treatment of the structure of abstract spaces, in particular linear, topological, metric, and normed spaces, as well as topological groups, in a rigorous and reader-friendly fashion. The assumed background knowledge on the part of the reader is modest, limited to basic concepts of finite dimensional linear spaces and elementary analysis. The book's aim is to serve as an introduction toward the theory of Hilbert spaces and the theory of operators.

The formalism of Hilbert spaces is fundamental to Physics, and in particular to Quantum Mechanics, requiring a certain amount of fluency with the techniques of linear algebra, metric space theory, and topology. Typical introductory level books devoted to Hilbert spaces assume a significant level of familiarity with the necessary background material and consequently present only a brief review of it. A reader who finds the overview insufficient is forced to consult other sources to fill the gap. Assuming only a rudimentary understanding of real analysis and linear algebra, this book offers an introduction to the mathematical prerequisites of Hilbert space theory in a single self-contained source. The final chapter is devoted to topological groups, offering a glimpse toward more advanced theory. The text is suitable for advanced undergraduate or introductory graduate courses for both Physics and Mathematics students.

The Structure of the Book

The book consists of six chapters, with an additional chapter of solved problems arranged by topic. Each chapter is composed of five sections, with each section accompanied by a set of exercises (with the exception of Chap. 6, which is shorter, and thus contains a single batch of exercises located at the end of the chapter). The total of 210 exercises and 50 solved problems comprise an integral part of the book designed to assist the reader, challenge her, and hone her intuition.

Chapter 1 contains a general introduction to real analysis and, in particular, to each of the subjects presented in the chapters that follow. Chapter 1 also contains a Preliminaries section, intended to quickly orient the reader as to the notation and concepts used throughout the book, starting with sets and ending with an axiomatic presentation of the real numbers.

Chapter 2 is devoted to linear spaces. At the advanced undergraduate level the reader is already familiar with at least some aspects of linear spaces, primarily finite dimensional ones. The chapter does not rely on any previous knowledge though, and is in that sense self-contained. However, the material is somewhat advanced since the focus is infinite dimensional linear spaces, which are technically relatively demanding.

Chapter 3 is an introduction to topology, a subject considered to be at a rather high level of abstraction. The main aim of the chapter is to familiarize the reader with the fundamentals of the theory, and in particular those that are most directly relevant for real analysis and Hilbert spaces. Care is taken to finding a reasonable balance between the study of extreme topology, i.e., spaces or phenomena that one may consider pathological but that hone the topological intuition, and mundane topology, i.e., those spaces or phenomena one is most likely to find in nature, but which may obscure the true nature of topology.

Chapter 4 is a study of metric spaces. Once the necessary fundamentals are covered, the main focus is complete metric spaces. In particular, the Banach Fixed-Point Theorem and Baire's Theorem are proved and completions are discussed, topics which are indispensable for Hilbert space theory.

Chapter 5 introduces and studies normed spaces and Banach spaces. Starting with semi-normed spaces the chapter establishes the fundamentals, and goes on to introduce Banach spaces, treating the Open Mapping Theorem, the Hahn-Banach Theorem, the Closed Graph Theorem, and, alluding to Hilbert space theory proper, the Riesz Representation Theorem.

Chapter 6 is a short introduction to topological groups, emphasizing their relation to Banach spaces. The chapter does not assume any knowledge of group theory, and thus, to remain self-contained, it presents all relevant group-theoretic notions. The chapter ends with a treatment of uniform spaces and a hint of their usefulness in the general theory.

A Word About the Intended Audience

The book is aimed at the advanced undergraduate or beginning postgraduate level, with the general prerequisite of sufficient mathematical maturity as expected at that level of studies. The book should be of interest to the student knowing nothing of

Hilbert space theory who wishes to master its prerequisites. The book should also be useful to the reader who is already familiar with some aspects of Hilbert space theory, linear spaces, topology, or metric spaces, as the book contains all the relevant definitions and pivotal theorems in each of the subjects it covers. Moreover, the book contains a chapter on topological groups and a treatment of uniform spaces, a topic usually considered at a more advanced level.

A Word About the Authors

The authors of the book, a physicist and a mathematician, by writing the entire book together, and through many arguments about notation and style, hope that this clash between the desires of a physicist to quickly yet intelligibly get to the point and the insistence of a mathematician on rigor and conciseness did not leave the pages of this book tainted with blood, but rather that it resulted in a welcoming introduction for both physicists and mathematicians interested in Hilbert space theory.

Prof. Carlo Alabiso obtained his Degree in Physics at Milan University, Italy, and then taught for more than 40 years at Parma University, Parma, Italy (with a period spent as a research fellow at the Stanford Linear Accelerator Center and at Cern, Geneva). His teaching encompassed topics in Quantum Mechanics, special relativity, field theory, elementary particle physics, mathematical physics, and functional analysis. His research fields include mathematical physics (Padé approximants), elementary particle physics (symmetries and quark models), and statistical physics (ergodic problems), and he has published articles in a wide range of national and international journals as well as the previous Springer book (with Alessandro Chiesa), Problemi di Meccanica Quantistica non Relativistica.

Dr. Ittay Weiss completed his B.Sc. and M.Sc. studies in Mathematics at the Hebrew University of Jerusalem and he obtained his Ph.D. in mathematics from Universiteit Utrecht in The Netherlands. He spent an additional 3 years in Utrecht as an assistant professor of mathematics where he taught mathematics courses across the entire undergraduate spectrum both at Utrecht University and at the affiliated University College Utrecht. He is currently a mathematics lecturer at the University of the South Pacific. His research interests lie in the fields of algebraic topology and operad theory, as well as the mathematical foundations of analysis and generalizations of metric spaces.

A Word of Gratitude

First and foremost, the authors extend their gratitude to Prof. Adriano Tomassini and to Dr. Daniele Angella for their contribution to the Solved Problems chapter. The authors wish to thank Rahel Berman for her numerous stylistic suggestions and to Alveen Chand for useful comments that improved the Preliminaries chapter.

Parma, Italy, June 2014 — Carlo Alabiso
Suva, Fiji — Ittay Weiss

Contents

Symbols

$\mathbb{N}$	The natural numbers
$\mathbb{Z}$	The integer numbers
$\mathbb{Q}$	The rational numbers
$\mathbb{R}$	The real numbers
$\mathbb{C}$	The complex numbers
id	Identity function
id_X	Identity function on the set X
I	Either the identity operator on a linear space or an interval
I_n	The $n \times n$ identity matrix
$\|X\|$	Cardinality of a set X
$\mathscr{P}(X)$	Power-set of the set X
$\in$	Set memberhood
$\subseteq$	Set inclusion
$\subset$	Proper set inclusion
$\cup$	Union of sets
$\cap$	Intersection of sets
$X - Y$	The relative complement of the set Y in the set X
$\sim$	Equivalence relation
$\sim_X$	Equivalence relation on the set X
$X/\sim$	Quotient set/space
$\cong$	Indicates isomorphic, isometric, or homeomorphic spaces
$\times$	The product of sets, linear spaces, topologies spaces, metric spaces, or Banach spaces
$\mathscr{B}$	Stands for a basis in a linear space or a topological space, or for a Banach space
$\mathscr{H}$	A Hilbert space
c	Space of convergent sequences of either real or complex numbers

c_0	Space of sequences of either real or complex numbers converging to 0
c_{00}	Space of eventually 0 sequences of either real or complex numbers
ℓ_p	Space of absolutely p-power summable sequences of either real or complex numbers
ℓ_∞	Space of bounded sequences of real or complex numbers
$C(I,\mathbb{R})$	Space of real-valued continuous functions on I
$C^k(I,\mathbb{R})$	Space of real-valued functions on I with a continuous k-th derivative
$C^\infty(I,\mathbb{R})$	Space of infinitely differentiable real-valued functions on I
$C(I,\mathbb{C})$	Space of complex-valued continuous functions on I
$\mathbb{R}^n$	Space of n-tuples of real numbers
$\mathbb{R}^\infty$	Space of infinite sequences of real numbers
$\mathbb{C}^n$	Space of n-tuples of complex numbers
$\mathbb{C}^\infty$	Space of infinite sequences of complex numbers
P_n	Space of polynomials of degree at most n with either real or complex coefficients
P	Space of all polynomials with either real of complex coefficients
K^B	Set of all functions from B to K
$(K^B)_0$	Space of all functions from a set B to a field K which are almost always 0
$\oplus$	Direct sum
$\\|x\\|$	Norm of a vector x
$\\|x\\|_p$	The ℓ_p or L_p norm of x
$\\|x\\|_\infty$	The ℓ_∞ or L_∞ norm of x
$\langle x, y\rangle$	The inner product of x and y
$p(x)$	The semi-norm of a vector x
Ker(T)	Kernel of a linear operator T
Im(T)	Image of a linear operator T
dim(V)	Dimension of a linear space V
τ	A topology
τ_X	A topology on the set X
$\mathbb{S}$	Sierpinski space
$\overline{S}$	Closure of a subset
$\partial(S)$	Boundary of a subset
int(S)	Interior of a subset
ext(S)	Exterior of subset
$B_\varepsilon(x)$	Open ball of radius ε and centre x
$\overline{B}_\varepsilon(x)$	Closed ball of radius ε and centre x
$\mathbb{R}_+$	The extended non-negative real numbers
d	Distance function in a metric space
d_X	Distance function on the set X
$\mathbf{B}(V,W)$	The set of bounded linear operators from V to W

$\mathrm{B}(\mathscr{B})$	The set of bounded linear operators on $\mathscr{B}$
V^*	The dual space of V
e	The identity element in a group

Chapter 1
Introduction and Preliminaries

Abstract This short chapter, consisting of two sections, contains a brief overview of the mathematical foundations of Hilbert spaces and analysis, a discussion of the real number system, and a detailed account of the preliminaries required for reading the rest of the book.

Keywords Set · Equivalence relation · Function · Zorn's Lemma · Relation · Cardinality · Countable set · Ordered set · Real numbers axioms · Set operations

1.1 Hilbert Space Theory—A Quick Overview

Hilbert space theory is the mathematical formalism for non-relativistic Quantum Mechanics. A Hilbert space is a linear space together with an inner product that endows the space with the structure of a complete metric space. Hilbert space theory is thus a fusion of algebra, topology, and geometry.

The reader is already familiar with the intermingling of algebra and geometry, namely in the linear space $\mathbb{R}^n$. Elements in $\mathbb{R}^n$ can be thought of as points in n-dimensional space but also as vectors. Typically, points have coordinates and vectors can be added and scaled. Moreover, in the presence of the standard inner product, given by $\langle x, y\rangle = \sum_{k=1}^{n} x_k y_k$, the length of a vector is given by the norm

$$\|x\| = \sqrt{\langle x, x\rangle}$$

and the angle between vectors can be computed by means of the formula

$$\theta = \arccos \frac{\langle x, y\rangle}{\|x\|\|y\|}.$$

At this level, the interaction between the algebra and the geometry is quite smooth. Things change though as soon as one considers infinite dimensional linear spaces, and this is also where topology comes into the picture. For finite dimensional linear spaces, bases (and thus coordinates) are readily available, all linear operators are continuous, convergence of operators has a single meaning, any linear space is

© Springer International Publishing Switzerland 2015
C. Alabiso and I. Weiss, *A Primer on Hilbert Space Theory*,
UNITEXT for Physics, DOI 10.1007/978-3-319-03713-4_1

naturally isomorphic to its double dual, and the closed unit ball is compact. These conveniences do not exist in the infinite dimensional case. While bases do exist, the proof of their existence is non-constructive and more often than not no basis can be given explicitly. Consequently, techniques that rely on coordinates and matrices are generally unsuitable. Linear operators need not be continuous, and in fact many linear operators of interest are not continuous. The space consisting of all linear operators between two linear spaces carries two different topologies, and thus two different notions of convergence. The correct notion of the dual space is that of all continuous linear operators into the ground field and even then the original space only embeds in its double dual. Lastly, the closed unit ball is not compact.

In a sense, the aim of Hilbert space theory is to develop adequate machinery to be able to reason about, and in, infinite dimensional linear spaces subject to these inherent difficulties and topological subtleties. The mathematics background required for the study of Hilbert spaces thus includes linear algebra, topology, the theory of metric spaces, and the theory of normed spaces. The theory of topological groups, a fusion between group theory and topology, arises naturally as well. Before we proceed to elaborate on each of these topics we visit the most fundamental space of them all, that of the real numbers.

1.1.1 The Real Numbers—Where it All Begins

Most fundamental to human observation of the outside world are the real numbers. The outcome of a measurement is almost always assumed to be a real number (at least in some ideal sense). That this idealization is deeply ingrained in the scientific lore is witnessed by the very name of the *real* numbers; among all numbers, these are the real ones! Whatever the reason may be for science's strong preference for the real numbers is a debate for philosophers. We merely observe that, whether the use of real numbers is justified or not, the success of the real numbers in science is unquestionable.

Mathematically though, the real numbers pose several non-trivial challenges. One of those challenges is the very definition of the real numbers, or, in other words, answering the question: what is a real number? For the ancient Greeks, roughly speaking until the Pythagoreans, the real numbers were taken for granted as being the same as what we would today call the rationals. That was the commonly held belief until the discovery that $\sqrt{2}$ is an irrational number (the precise circumstances of that discovery are unclear). The construction of precise models for the reals had to wait numerous centuries, and so did the discovery of transcendental numbers and the understanding of just how many real numbers there are.

1.1.1.1 Constructing the Real Numbers

The reader is likely to have her own personal idea(lization) regarding what the real numbers really are. Any precise interpretation or elucidation of the true entities which are the real numbers belongs to philosophy. Choosing to stay firmly afoot in mathematics, we present now one of many precise constructions of a model of the real numbers. As with the construction of any model, one can not construct something from nothing. We thus assume the reader accepts the existence of (a model of) the rational numbers. In other words, we pretend to agree on what the rational numbers are, but to have no knowledge of the real number system. We will give the precise construction but, to remain in the spirit of an introduction, we avoid the details of the proofs.

To motivate the following construction of the real numbers from the rational numbers, recall that the rational numbers are dense in the real numbers, meaning that between any two real numbers there exists a rational number. This simple observation gives rise to the crucial fact that every real number is the limit of a sequence of rational numbers. Thus, the convergent sequences of rational numbers give us access to all of the real numbers. Since many different sequences of rational numbers may converge to the same real number, the set of all convergent sequences of rational numbers is too wasteful to be taken as the definition of the real numbers. One needs to introduce an equivalence relation on it, one that identifies two sequences of rational numbers if they converge to the same real number.

To conclude the discussion so far, if S denotes the set of all convergent sequences of rational numbers, then we expect to be able to identify an equivalence relation $\sim$ on S such that $S/\sim$ forms a model of the real numbers. However, to even take the first step in this plan without committing the sin of a circular argument, one must first be able to identify the convergent sequences among all sequences of rational numbers without any a-priori knowledge of the real numbers. That task is achieved by appealing to the familiar notion of a Cauchy sequence, a condition which, for sequences of real numbers, is well-known to be equivalent to convergence. The key observation is that the Cauchy condition for sequences of rational numbers can be stated without mentioning real numbers at all. In this way the convergent sequences of rational numbers are carved from the set of all sequences of rationals by means of an inherently rational criterion.

The construction we give is due to Georg Cantor, the creator of the theory of sets. On the set $\mathbb{Q}$ of rational numbers, which we assume is endowed with the familiar notions of addition and multiplication, and thus also with the notions of subtraction and division, consider the function $d : \mathbb{Q} \times \mathbb{Q} \to \mathbb{Q}$ given by $d(x, y) = |x - y|$. We now define a Cauchy sequence to be a sequence (x_n) of rational numbers which satisfies the condition that for every rational number $\varepsilon > 0$ there exists $N \in \mathbb{N}$ such that $d(x_n, x_m) < \varepsilon$ for all $n, m > N$. Let now S be the set of all Cauchy sequences and define the relation $\sim$ on S as follows. Given Cauchy sequences $(x_n), (y_n) \in S$, declare that $(x_n) \sim (y_n)$ if $d(x_n, y_n) \to 0$. Note that convergence to 0 (or any other rational number) of a sequence of rational numbers can be stated without assuming the existence of real numbers. Indeed, the meaning of $(x_n) \sim (y_n)$ is that for every rational number $\varepsilon > 0$ there exists an $N \in \mathbb{N}$ such that $d(x_n, y_n) < \varepsilon$ for all $n > N$.

It is a straightforward matter to show that $\sim$ is an equivalence relation on S. We now define the set of real numbers to be $\mathbb{R} = S/\sim$, namely the quotient set determined by the equivalence relation. One immediate thing to notice is that $\mathbb{R}$ contains a copy of $\mathbb{Q}$. Indeed, for any rational number $q \in \mathbb{Q}$ consider the constant sequence $x_q = (x_n = q)$. It is clearly a Cauchy sequence, and thus $x_q \in S$, and its equivalence class $[x_q]$ is thus, per definition, a real number. Further, if $q' \neq q$ is another rational, then $[x_q] \neq [x_{q'}]$ since the respective constant sequences are not equivalent, and thus not identified in the quotient set. In other words, the function $q \mapsto [x_q]$ is an injection from $\mathbb{Q}$ to $\mathbb{R}$.

The familiar algebraic structure of $\mathbb{Q}$ carries over to $\mathbb{R}$ as follows. Given real numbers x, y, represented by Cauchy sequences (x_n) and (y_n) respectively, the sum $x + y$ is defined to be represented by $(x_n + y_n)$ and the product xy to be represented by $(x_n y_n)$. Of course, one needs to check these are well-defined notions, i.e., that the proposed sequences are Cauchy and that the resulting sum and product are independent of the choice of representatives. This verification, especially for the sum, is straightforward. Once that is done, the verification of the familiar algebraic properties of addition and multiplication is easily performed, proving the $\mathbb{R}$ is a field. Similarly, the order structure of $\mathbb{Q}$ carries over to $\mathbb{R}$ by defining, for real numbers x and y as above, that $x \leq y$ if either $x = y$ or if $x_n \leq y_n$ for all but finitely many n. It is quite straightforward to verify that $\mathbb{R}$ is then an ordered field. Finally, and by far least trivially, $\mathbb{R}$ satisfies the least upper bound property, proving that it is Dedekind complete.

The embedding of $\mathbb{Q}$ in $\mathbb{R}$ described above is easily seen to respect the order and algebraic structure of $\mathbb{Q}$. Therefore the field $\mathbb{Q}$ of rational numbers we started with can be identified with its isomorphic copy in $\mathbb{R}$, namely the image of the embedding $\mathbb{Q} \to \mathbb{R}$, and one then considers $\mathbb{R}$ as an extension of $\mathbb{Q}$.

1.1.1.2 Transcendentals and Uncountablity

The divide of the real numbers into rational and irrational numbers represents the first fundamental indication of the complexity of the real numbers. Together with the realization that $\sqrt{2}$ is an irrational number comes the necessity, if only out of curiosity, of asking about the nature of other real numbers, for instance π and e. The number e was introduced by Jacob Bernoulli in 1683 but it was only some 50 years later that Leonard Euler established its irrationality. In sharp contrast, the number π, known since antiquity, was only shown to be irrational in 1761 by Johann Heinrich Lambert. If this is not indication enough that the real number system holds more secrets than one would superficially expect, the fact that to date it is unknown whether or not $\pi + e$ is irrational should remove any doubt.

Some real numbers, while irrational, may still be nearly rational, in the following sense. Consider a rational number $r = p/q$. Clearly, r satisfies the equation

$$qt - p = 0,$$

and thus r is a root of a polynomial with integer coefficients. More generally, any real number which is a root of a polynomial with integer coefficients is called an algebraic number. For instance, $\sqrt{2}$ is algebraic since it is a root of the polynomial $t^2 - 2$. A real number that is not algebraic is called transcendental. The existence of transcendental numbers was not established until 1844 by Joseph Liouville who, shortly afterwards, constructed the first explicit example of a transcendental number (albeit a somewhat artificial one). More prominent members of the transcendental class are the number e, proved transcendental by Charles Hermite in 1873, and the number π, proved transcendental by Ferdinand von Lindemann in 1882. While much is known about transcendental numbers, it is still unknown if $\pi + e$ is algebraic or not.

In 1884 Georg Cantor, following shortly after Liouville's proof of the existence of transcendental numbers, proved not only that transcendental numbers exist but also that in some sense the vast majority of all real numbers are transcendental. The technique employed by Cantor was that of transfinite counting. With his notion of cardinality of sets, Cantor counted how many algebraic numbers there are and how many real numbers there are, and demonstrated that there are strictly more real numbers than algebraic ones. The inevitable conclusion that transcendental numbers exist, a conclusion that was met with some resistance, forces us to face the following equally inevitable truth. Similarly to Cantor's counting of the algebraic numbers one can also count all sentences in any given natural language, for instance English. As it turns out, the cardinality of all potential English descriptions of real numbers is the same as the cardinality of the algebraic numbers, and thus is strictly smaller than the cardinality of all real numbers. We must now conclude that there exist real numbers that can not ever, even potentially, be described in any way at all.

We now end this short historical excursion and turn to quickly address the main topics covered in this book.

1.1.2 Linear Spaces

A linear space, also known as a vector space, embodies what is perhaps the simplest notion of a mathematical space. When thought of as modeling actual physical space, a linear space appears to be the same in all directions and is completely devoid of any curvature. Linear spaces, and the closely related notion of linear transformation or linear operator, are fundamental objects. For instance, the derivative of a function at a point is best understood as a linear operator on tangent spaces, particularly for functions of several variables. Differentiable manifolds are spaces which may be very complicated but, at each point, they are locally linear. It is perhaps somewhat of an exaggeration, but it is famously held to be true that if a problem can be linearized, then it is as good as solved.

1.1.3 Topological Spaces

Topology is easy to define but hard to explain. The ideas that led to the development of topology were lurking beneath the surface for some time and it is difficult to pinpoint the exact time in history when topology was born. What is clear is that after its birth the advancement of topology in the first half of the 20th century was rapid. The unifying power of topology is immense and its explanatory ability is a powerful one. For instance, familiar theorems of first year real analysis, such as the existence of maxima and minima for continuous functions on a closed interval, or the uniform continuity of a continuous function on a closed interval, may make the closed interval appear to have special significance. Topology is able to clarify the situation, identifying a particular topological property of the closed interval, namely that it is compact, as the key ingredient enabling the proof to carry over into much more general situations.

1.1.4 Metric Spaces

Metric spaces appeared in 1906 in the Ph.D dissertation of Maurice Réne Fréchet. A metric space is a set where one can measure distances between points, and in a strong sense a metric space carries much more geometric information than can be described by a topological space. With stronger axioms come stronger theorems but also less examples. However, the axioms of a metric space allow for a vast and varied array of examples and the theorems one can prove are very strong. In particular, complete metric spaces, i.e., those that, intuitively, have no holes, admit two very strong theorems. One is the Banach Fixed-Point Theorem and the second is Baire's Theorem. The former can be used to solve, among other things, differential equations, while the latter has deep consequences to the structure of complete metric spaces and to continuous functions.

1.1.5 Normed Spaces and Banach Spaces

A very basic attribute one can associate with a vector is its norm, i.e., its length. The abstract formalism is given by the axioms of a normed space. The presence of a norm allows one to define the distance between any two vectors, by the so-called induced metric, and one obtains a metric space, as well as the topology it induces. Thus any normed space immediately incorporates both algebra and geometry. A Banach space is a normed space which, as a metric space, is complete. The interaction between the algebra and the geometry is then particularly powerful, allowing for very strong results.

1.1.6 Topological Groups

Banach spaces are a fusion of algebra and geometry. Similarly, topological groups are a fusion between algebra and topology. Unlike a Banach space, the algebraic structure is that of a group while the geometry is reduced from a metric to a topology. Thus, a topological group is a much weaker structure than a Banach space, yet the interaction between the algebra and the topology still gives rise to a very rich and interesting theory.

1.2 Preliminaries

The aim of this section is to recount some of the fundamental notions of set theory, the common language for rigorous mathematical discussions. It serves to quickly orient the reader with respect to the notation used throughout the book as well as to present fundamental results on cardinals and demonstrate the technique of Zorn's Lemma. This section is designed to be skimmed through and only consulted for a more detailed reading if needed further on. For that reason, and unlike the rest of the book, the style of presentation in this section is rather condensed.

1.2.1 Sets

The notion of a *set* is taken to be a primitive notion, left undefined. In an axiomatic approach to sets one lists certain well formulated and carefully chosen axioms, while in a naive approach one relies on a common understanding of what sets are, avoiding the technical difficulties of precise definitions at the cost of some rigor. We adopt the naive approach and thus for us a set is a *collection* of *elements*, where no repetitions are allowed and no ordering of the elements of the set is assumed.

1.2.2 Common Sets

Among the sets we will encounter are the sets $\mathbb{N} = \{1, 2, 3, \ldots\}$ of natural numbers, $\mathbb{Z} = \{\ldots, -3, -2, -1, 0, 1, 2, 3, \ldots\}$ of integers, $\mathbb{Q} = \{p/q \mid p, q \in \mathbb{Z}, q \neq 0\}$ of rational numbers, $\mathbb{R}$ of real numbers, and $\mathbb{C} = \{x + iy \mid x, y \in \mathbb{R}\}$ of complex numbers. The notation illustrates two commonly occurring ways for introducing sets, the informal $\ldots$ which is meant to indicate that the reader knows what the author had in mind, and the precise form $\{x \in X \mid P(x)\}$, to be interpreted as the set of all those elements x in X satisfying property P. The *empty set* is denoted by $\emptyset$, and it simply has no elements in it. A set $\{x\}$, consisting of just a single element, is called a *singleton* set.

1.2.3 Relations Between Sets

The notation $x \in X$ was already used above to indicate that x is a *member* of the set X. To indicate that x is not a member of X we write $x \notin X$. Two sets X and Y may have the property that any element $x \in X$ is also a member of Y, a situation denoted by $X \subseteq Y$, meaning that X is *contained* in Y, X is a *subset* of Y, Y *contains* X, and that Y is a *superset* of X. If moreover $X \neq Y$, then X is said to be a *proper subset* of Y, denoted by $X \subset Y$. In particular, $X \subset Y$ means that every element $x \in X$ satisfies $x \in Y$ and that for at least one element $y \in Y$ it holds that $y \notin X$. An often useful observation is that two sets X and Y are equal, denoted by $X = Y$, if, and only if, $X \subseteq Y$ and $Y \subseteq X$.

1.2.4 Families of Sets; Union and Intersection

Any set I may serve as an indexing set for a set of sets. In that case it is customary to speak of a *family of sets*, a *collection of sets*, or an *indexed set*. For instance, consider $I = [0, 1] = \{t \in \mathbb{R} \mid 0 \leq t \leq 1\}$, and for each $t \in I$ let $X_t = (0, t) = \{x \in \mathbb{R} \mid 0 < x < t\}$. Then $\{X_t\}_{t \in I}$ is a family of sets indexed by I. For any such indexed collection, its *union* is the set

$$\bigcup_{i \in I} X_i$$

which is the set of all those elements x that belong to X_i for some $i \in I$ (that may depend on x). Moreover, if $X_i \cap X_j = \emptyset$ for all $i, j \in I$ with $i \neq j$, then the collection is said to be *pairwise disjoint* and the union is then said to be a *disjoint union*. The *intersection* of the collection is the set

$$\bigcap_{i \in I} X_i$$

which is the set of all those elements x that belong to X_i for all $i \in I$. For the indexed collection given above one may verify that

$$\bigcup_{t \in I} X_t = (0, 1)$$

and that

$$\bigcap_{t \in I} X_t = \emptyset.$$

When the indexing set is finite, one typically writes

$$X_1 \cup X_2 \cup \cdots \cup X_n$$

and

$$X_1 \cap X_2 \cap \cdots \cap X_n$$

for the union and intersection, respectively.

1.2.5 Set Difference, Complementation, and De Morgan's Laws

The *set difference* $X - Y$ is the set $\{x \in X \mid x \notin Y\}$. In the presence of a *universal set* or a *set of discourse*, that is a set U which is the relevant ambient set in a given context, the *complement* of a set $X \subseteq U$ is the set $X^c = \{y \in U \mid y \notin X\}$. In other words, $X^c = U - X$. Needless to say, the notation X^c pre-supposes knowledge of U, and different choices of U yield different complements. In this context we mention that the set difference $X - Y$ is also known as the *relative complement* of X in Y.

Among the many properties of sets and the operations of union, intersection, and complementation we only mention *De Morgan's laws*: given any indexed collection $\{X_i\}_{i \in I}$ of subsets of a set X, the equalities

$$(\bigcup_{i \in I} X_i)^c = \bigcap_{i \in I} X_i^c$$

and

$$(\bigcap_{i \in I} X_i)^c = \bigcup_{i \in I} X_i^c$$

both hold. Here the complements are relative to the set X.

1.2.6 Finite Cartesian Products

Given m sets ($m \geq 1$), $X_1, \ldots, X_m$, their *cartesian product* is the set

$$X_1 \times \cdots \times X_m = \{(x_1, \ldots, x_m) \mid x_k \in X_k, \text{ for all } 1 \leq k \leq m\}.$$

In other words, the cartesian product of m sets $X_1, \ldots, X_m$ (in that order) is the set consisting of all m-tuples where the k-th component (for $1 \leq k \leq m$) is taken from the k-th set. In case $X_k = X$ for all $1 \leq k \leq m$, the cartesian product is denoted by X^m and is called the *m-fold cartesian product* of X with itself.

1.2.7 Functions

Given any two sets X and Y, a *function* or a *mapping*, denoted by $f : X \to Y$or $X \xrightarrow{f} Y$ is, informally, a rule that associates with each element $x \in X$ an element $f(x) \in Y$. More formally, a function $f : X \to Y$ is a subset $f \subseteq X \times Y$ satisfying the property that for all $x \in X$ and $y_1, y_2 \in Y$, if $(x, y_1) \in f$ and $(x, y_2) \in f$, then $y_1 = y_2$, and that for all $x \in X$ there exists some $y \in Y$ with $(x, y) \in f$. The set X is the *domain* of the function while Y is the *codomain*. For any set X there is always the *identity function* $\mathrm{id}_X : X \to X$ given by $\mathrm{id}_X(x) = x$. Further, given functions $f : X \to Y$ and $g : Y \to Z$, their *composition* is defined to be the function $g \circ f : X \to Z$ given by $(g \circ f)(x) = g(f(x))$. It is obvious that composition of functions is associative, that is, if

$$X_1 \xrightarrow{f} X_2 \xrightarrow{g} X_3 \xrightarrow{h} X_4,$$

then

$$h \circ (g \circ f) = (h \circ g) \circ f.$$

A function $f : X \to Y$ is *injective* or *one-to-one* if $f(x_1) = f(x_2)$ implies $x_1 = x_2$, for all $x_1, x_2 \in X$. A function $f : X \to Y$ is *surjective* or *onto* if for all $y \in Y$ there exists an $x \in X$ such that $f(x) = y$. A function $f : X \to Y$ is called *bijective* or *invertible* if it is both injective and surjective. Equivalently, a function $f : X \to Y$ is invertible if there exists a function $f^{-1} : Y \to X$, called the *inverse* of f, such that $f \circ f^{-1} = \mathrm{id}_Y$ and $f^{-1} \circ f = \mathrm{id}_X$. Such an inverse, if it exists, is easily seen to be unique.

Given a function $f : X \to Y$ and a subset $S \subseteq X$, the *restriction* of f to S is the function $g : S \to Y$ given by $g(s) = f(s)$. This restriction function is denoted by $f|_S$. We also mention that any two sets X and Y with $X \subseteq Y$ give rise to the *inclusion function* $\iota : X \to Y$, defined by $\iota(x) = x$. Note that the inclusion function differs from the identity function only by the codomain, not by the functional assignments.

1.2.8 Arbitrary Cartesian Products

The *cartesian product* of any indexed family $\{X_i\}_{i\in I}$ of sets is the set

$$\{f : I \to \bigcup_{i\in I} X_i \mid f(i) \in X_i,\ \forall i \in I\}.$$

To justify the definition, observe that when $I = \{1, 2, 3, \ldots, m\}$, each function f as above can be uniquely identified with the m-tuple $(f(1), f(2), \ldots, f(m))$, in other words, with an element of $X_1 \times \cdots \times X_m$, so this definition recovers the definition of the cartesian product of finitely many sets, as given above. When $X_i = X$ for all $i \in I$, the cartesian product is denoted by X^I. Notice that X^I is simply the set of all functions $f : I \to X$.

1.2.9 Direct and Inverse Images

The *power set* of a set X is the set of all subsets of X, and is denoted by $\mathscr{P}(X)$. Every function $f : X \to Y$ induces the *direct image* function $f : \mathscr{P}(X) \to \mathscr{P}(Y)$, denoted again by f, given by $f(A) = \{f(a) \mid a \in A\}$, for all $A \subseteq X$ (notice that this definition introduces a new way to construct sets). Similarly, any function $f : X \to Y$ induces the *inverse image* function $f^{-1} : \mathscr{P}(Y) \to \mathscr{P}(X)$, which, for every $B \subseteq Y$, is given by $f^{-1}(B) = \{x \in X \mid f(x) \in B\}$. Notice that $f^{-1}(Y) = X$ but that the inclusion $f(X) \subseteq Y$ may be proper. The set $f(X)$ is called the *image* of f. Note that $f(X) = Y$ if, and only if, f is surjective. We remark that the notation f^{-1} for the inverse image conflicts with the notation for the inverse function of f (when it exists) and thus some care should be exercised.

The following properties of functions are easily verified. If $X \xrightarrow{f} Y \xrightarrow{g} Z$ are functions and $S \subseteq Z$ is a subset, then

$$(g \circ f)^{-1}(S) = f^{-1}(g^{-1}(S)).$$

Further, for all subsets $S_1, S_2 \subseteq Y$,

$$f^{-1}(S_1 \cap S_2) = f^{-1}(S_1) \cap f^{-1}(S_2)$$

and

$$f^{-1}(S_1 \cup S_2) = f^{-1}(S_1) \cup f^{-1}(S_2).$$

More generally, given any collection $\{S_i\}_{i\in I}$ of subsets of Y,

$$f^{-1}(\bigcap_{i\in I} S_i) = \bigcap_{i\in I} f^{-1}(S_i)$$

and

$$f^{-1}(\bigcup_{i\in I} S_i) = \bigcup_{i\in I} f^{-1}(S_i).$$

1.2.10 Indicator Functions

Let $\mathbb{S}$ be the set $\{0, 1\}$ and fix some set X. The subsets of X are easily seen to be in bijective correspondence with the functions $X \to \mathbb{S}$. The correspondence is given as follows. Given a subset $S \subseteq X$ let $F(S) : X \to \mathbb{S}$ be given by

$$F(S)(x) = \begin{cases} 1 & \text{if } x \in S \\ 0 & \text{if } x \notin S. \end{cases}$$

The function $F(S)$ is called the *indicator function* of the subset $S \subseteq X$. Conversely, given any function $f : X \to \mathbb{S}$, the subset of X it determines is $G(f) = f^{-1}(\{1\})$. It is easily verified that F and G are each other's inverses, setting up the stated bijective correspondence. Note that $f^{-1}(\{0\})$ is the complement $X - f^{-1}(\{1\})$.

1.2.11 Cardinality

Two sets X and Y are said to have the same *cardinality* if there exists a bijection $f : X \to Y$, in which case we write $|X| = |Y|$. If there is an injection $f : X \to Y$, then X is said to have cardinality less than or equal to that of Y, and we write $|X| \leq |Y|$.

Two finite sets have the same cardinality if, and only if, they have the same number of elements. It is thus common to write $|X|$ for a finite set X to indicate the number of elements in X.

A set X is said to be *countable* if it is empty or if there exists a surjective function $f : \mathbb{N} \to X$. Equivalently, a set is countable if it has the same cardinality as some subset of $\mathbb{N}$. Equivalently still, a set X is countable precisely when there exists an injective function $X \to \mathbb{N}$. A set that is both countable and infinite is said to be *countably infinite*. A set that is not countable is called *uncountable*. A set which has the same cardinality as $\mathbb{R}$ is said to have the cardinality of the *continuum*. Examples of countable sets include every finite set, the set $\mathbb{N}$, as well as the set $\mathbb{Z}$ of integers and the set $\mathbb{Q}$ of rational numbers. Sets of the cardinality of the continuum include the set $\mathbb{R}$ of real numbers as well as $\mathbb{C}$ of complex number, the n-fold products $\mathbb{R}^n$ and $\mathbb{C}^n$ for any $n \in \mathbb{N}$, as well as the power-set $\mathscr{P}(\mathbb{N})$, namely the set of all subsets of natural numbers, and the set of all continuous functions $f : \mathbb{R} \to \mathbb{R}$. Uncountable sets of cardinality larger than that of the continuum include the power-set $\mathscr{P}(\mathbb{R})$ of all subsets of real numbers, as well as the set of all functions $f : \mathbb{R} \to \mathbb{R}$.

1.2.12 The Cantor-Shröder-Bernstein Theorem

The result we present now is a very convenient tool in establishing that two sets have the same cardinality.

Theorem 1.1 (Cantor–Shröder–Bernstein) *For all sets X and Y, if $|X| \leq |Y|$ and $|Y| \leq |X|$, then $|X| = |Y|$.*

Proof By the condition in the assertion, there exists an injective function $f : X \to Y$ and an injective function $g : Y \to X$. To construct a bijection $h : X \to Y$, we consider the behaviour of elements in both X and Y with respect to the given functions f and g. Notice that since g is injective, the inverse image $g^{-1}(x_0)$, for any $x_0 \in X$, is either empty or is of the form $\{y_0\}$, where $g(y_0) = x_0$. Similarly, $f^{-1}(y_1)$, for any $y_1 \in Y$, is either empty or is of the form $\{x_1\}$ with $f(x_1) = y_1$. We thus write $f^{-1}(y_1) = x_1$ and $g^{-1}(x_0) = y_0$, when these sets are not empty. Starting with $x_0 \in X$ we may successively consider the sequence

$$x_0 \mapsto g^{-1}(x_0) = y_0 \mapsto f^{-1}(y_0) = x_1 \mapsto g^{-1}(x_1) = y_1 \mapsto \cdots$$

where at each step we apply f^{-1} or g^{-1} alternatively, if they exist. Such a sequence exhibits precisely one of three possible behaviours. It may be infinite, in which case x_0 is said to have type ∞, or it may terminate since $g^{-1}(x_k)$ was empty, in which case x_0 is said to have type X, or the sequence may stop due to the fact that $f^{-1}(y_k)$ was empty, in which case x_0 is said to have type Y. In exactly the same way, each element of Y can be classified as having precisely one type, either type ∞, or type X or type Y.

Let us now denote by X_∞ all type ∞ elements in X, by X_X all type X elements in X, and by X_Y all type Y elements in X. Similarly, introduce the notation Y_∞, Y_X, and Y_Y. Thus

$$X = X_\infty \cup X_X \cup X_Y$$

and the union is pairwise disjoint. Similarly we obtain the pairwise disjoint union

$$Y = Y_\infty \cup Y_X \cup Y_Y.$$

The function given by

$$h(x) = \begin{cases} f(x) & \text{if } x \in X_X \\ g^{-1}(x) & \text{if } x \in X_Y \\ f(x) & \text{if } x \in X_\infty \end{cases}$$

is now easily seen to be a bijection, for instance by constructing an inverse function for it. □

1.2.13 Countable Arithmetic

Theorem 1.2 *A subset of a countable set is countable.*

Proof Suppose a set X is countable and $S \subseteq X$ is a subset. If $S = \emptyset$, then it is countable, so we may assume $S \neq \emptyset$ and let us fix some $s_0 \in S$. Since X is countable, there exists a surjection $f : \mathbb{N} \to X$. Clearly, there is a surjection $g : X \to S$ (for instance, $g(x) = x$ if $x \in S$, and $g(x) = s_0$ if $x \notin S$) and, since the composition of surjective functions is surjective, the function $g \circ f : \mathbb{N} \to S$ is surjective, showing that S is countable. □

Theorem 1.3 *The cartesian product of finitely many countable sets is countable.*

Proof Let $X_1, \ldots, X_m$ be countable sets and choose, for each $1 \leq k \leq m$, an injective function $f_k : X_k \to \mathbb{N}$. These functions together give rise to

$$f : X_1 \times \cdots \times X_m \to \mathbb{N}^m$$

given by

$$f(x_1, \ldots, x_m) = (f_1(x_1), \ldots, f_m(x_m)),$$

which is clearly injective. Now fix m distinct prime numbers $p_1, \ldots, p_m$ and let

$$g : \mathbb{N}^m \to \mathbb{N}$$

be given by

$$g(n_1, \ldots, n_m) = p_1^{n_1} \cdot \cdots \cdot p_m^{n_m}.$$

The fact that every natural number has an essentially unique decomposition into prime factors implies that g is injective. To conclude then, the composition

$$X_1 \times \cdots \times X_m \xrightarrow{f} \mathbb{N}^m \xrightarrow{g} \mathbb{N}$$

is injective, and thus $X_1 \times \cdots \times X_m$ is countable. □

Theorem 1.4 *A countable union of countable sets is itself countable.*

Proof Let $\{X_m\}_{m\in\mathbb{N}}$ be a countable family of countable sets. We may assume, without loss of generality, that the collection is pairwise disjoint and that $X_m \neq \emptyset$ for all $m \in \mathbb{N}$. Since each X_m is countable, there exists a surjective function $f_m : \mathbb{N} \to X_m$. Let

$$X = \bigcup_{m\in\mathbb{N}} X_m$$

and define now the function

$$g : \mathbb{N} \times \mathbb{N} \to X$$

by

$$g(m, n) = f_m(n),$$

which is clearly surjective. To finish the proof, note that $\mathbb{N} \times \mathbb{N}$, being the product of two countable sets, is countable, and thus there is a surjection $h : \mathbb{N} \to \mathbb{N} \times \mathbb{N}$. The composition

$$\mathbb{N} \xrightarrow{h} \mathbb{N} \times \mathbb{N} \xrightarrow{f} X$$

is thus a surjection, proving that X is countable. □

1.2.14 Relations

A *relation* R on a set X is a subset $R \subseteq X \times X$, but we write xRy instead of $(x, y) \in R$. The relation R is said to be *reflexive* if xRx holds for all $x \in X$. The relation R is *symmetric* if xRy implies yRx, for all $x, y \in X$. The relation R is *anti-symmetric* if, for all $x, y \in X$, the assertions xRy and yRx together imply that $x = y$. Finally, the relation R is said to be *transitive* if xRy and yRz together imply xRz, for all $x, y, z \in X$.

1.2.15 Equivalence Relations

A relation R on a set X is called an *equivalence relation* if it is reflexive, symmetric, and transitive, in which case we denote R by $\sim$, and thus write $x \sim y$ instead of xRy. If $x \sim y$ we say that x and y are *equivalent*. A *partition* of X is a collection $\{X_i\}_{i\in I}$ of non-empty pairwise disjoint subsets of X such that $\bigcup_{i\in I} X_i = X$.

Given a partition $\{X_i\}_{i \in I}$ of X, defining $x \sim y$ precisely when there exists an $i \in I$ with $x, y \in X_i$ is easily seen to define an equivalence relation. If we denote by $\mathrm{Par}(X)$ the set of all partitions of X and by $\mathrm{Equ}(X)$ the set of all equivalence relations on X, then the process just described defines a function $E : \mathrm{Par}(X) \to \mathrm{Equ}(X)$. Conversely, given an equivalence relation $\sim$ on X we may define, for each $x \in X$, the set

$$[x] = \{y \in X \mid x \sim y\}$$

of all those elements $y \in X$ which are equivalent to x. This set is known as the *equivalence class* of x, and x is called a *representative* of the equivalence class $[x]$ (if it is important to emphasize the relation $\sim$, then we write $[x]_\sim$). It can easily be verified that two elements $x, y \in X$ represent the same equivalence class, that is $[x] = [y]$, if, and only if, $x \sim y$. It is completely straightforward to demonstrate that the set $\{[x] \mid x \in X\}$ of all equivalence classes is a partition of X, and we thus obtain a function $P : \mathrm{Equ}(X) \to \mathrm{Par}(X)$. It is quite easy to verify that in fact P is the inverse function of E and thus we have established a bijective correspondence between equivalence classes on X and partitions of X.

Given an equivalence relation $\sim$ on a set X, the set $\{[x] \mid x \in X\}$ is denoted by $X/{\sim}$ and is called the *quotient set* of X *modulo* $\sim$. There is also the corresponding function $\pi : X \to X/{\sim}$, given by $\pi(x) = [x]$, called the *canonical projection.*

1.2.16 Ordered Sets

A *poset* (short for *partially ordered set*) is a pair $(X, \leq)$ where X is a set and $\leq$ is a relation on X which is reflexive, transitive, and anti-symmetric.

Example 1.1 Given any set X, let $\mathscr{P}(X)$ be the set of all subsets of X. Defining, for subsets $Y_1, Y_2 \subseteq X$, that $Y_1 \leq Y_2$ precisely when $Y_1 \subseteq Y_2$, is easily seen to endow $\mathscr{P}(X)$ with the structure of a poset. In this case we say that $\mathscr{P}(X)$ is ordered by *set inclusion*. More generally, if A is some collection of subsets of X, then it too can be endowed with the ordering induced by set inclusion, in precisely the same manner. A related construction is to consider the set P of all pairs (Y, f) where $Y \subseteq X$ and $f : Y \to Z$ is a function to some fixed set Z. Defining $(Y_1, f_{1,}) \leq (Y_2, f_2)$ when $Y_1 \subseteq Y_2$ and when f_2 extends f_1 (which means that $f_2(y) = f_1(y)$ holds for all $y \in Y_1$) is again easily seen to endow P with a poset structure. Again, one may consider variants of this construction, for instance by demanding extra conditions on the sets Y or on the functions f.

In the context of a general poset, the meaning of $x < y$ is taken to be: $x \leq y$ and $x \neq y$. The relations $x > y$ and $x \geq y$ are similarly derived from the given relation $\leq$. Since not all elements x and y in a poset need be *comparable* (x and y are comparable when either $x \leq y$ or $y \leq x$), the correct interpretation of, for instance, the negation of $x \leq y$ is not that $x > y$ but rather that either x and y are incomparable, or $x > y$.

1.2.17 Zorn's Lemma

A poset P is said to be a *total order* or *linearly ordered* if for all $x, y \in P$ either $x \leq y$ or $y \leq x$ holds. The poset $\mathscr{P}(X)$ discussed in Example 1.1 is linearly ordered if, and only if, $|X| \leq 1$.

Any subset $S \subseteq P$ in a poset $(P, \leq)$ inherits a poset structure from P. We also say that P induces a poset structure on S. In more detail, the poset S is given by the ordering $\preceq$ defined, for all $x, y \in S$, by $x \preceq y$ precisely when $x \leq y$ in P. We usually do not make any notational distinction between $\leq$ and the *induced order* $\preceq$, and thus simply write $x \leq y$ when referring to the ordering in S.

Definition 1.1 A *chain* in a poset P is a subset $S \subseteq P$ which, with the induced ordering, is linearly ordered. An *upper bound* of a set $S \subseteq P$ (be it a chain or not) is an element $y \in P$ such that $x \leq y$ holds for all $x \in S$. A *maximal* element in a poset P is an element y_M such that $y_M < x$ does not hold for any $x \in P$.

Lemma 1.1 (Zorn's Lemma) *Let P be a non-empty poset. If every chain in P has an upper bound, then P has a maximal element.*

Zorn's Lemma is a powerful result that is more of a proof technique than a lemma, in much the same way as proof by induction is a proof technique rather than a lemma. A proof of Zorn's Lemma is subtle and in fact it is well-known that Zorn's Lemma is equivalent to the Axiom of Choice.

1.2.18 A Typical Application of Zorn's Lemma

Given any two sets X and Y, it is natural to wonder if their cardinalities compare. That is, is it always the case that either $|X| \leq |Y|$, or $|Y| \leq |X|$? A positive answer to this question turns out to be equivalent to the Axiom of Choice. We will only prove half of this equivalence, primarily to illustrate a typical application of Zorn's Lemma.

Theorem 1.5 *For all sets X and Y, either $|X| \leq |Y|$ or $|Y| \leq |X|$.*

Proof The case where either X or Y is empty is easily dispensed with, and so we proceed under the assumption that both are non-empty. As a first step we construct a suitable poset, similar to the one given in Example 1.1.

Let P be the set of triples (X', f, Y') where $X' \subseteq X$, $Y' \subseteq Y$, and $f : X' \to Y'$ is a bijective function. The partial order structure on P is given by

$$(X', f, Y') \leq (X'', g, Y'')$$

precisely when

$$X' \subseteq X''$$
$$Y' \subseteq Y''$$
$$g(x) = f(x) \text{ for all } x \in X'.$$

It is straightforward to verify that this is indeed a partial order.

Next, we show that $(P, \leq)$ satisfies the conditions of Zorn's Lemma. Firstly, $P \neq \emptyset$ since the triple $(\emptyset, \emptyset, \emptyset)$ is always in P. Next, suppose that $\{(X_i, f_i, Y_i)\}_{i \in I}$ is a chain in P, and we shall construct an upper bound for it. Let

$$X_0 = \bigcup_{i \in I} X_i$$
$$Y_0 = \bigcup_{i \in I} Y_i$$

and $f : X_0 \to Y_0$ given by

$$f(x) = f_{i_x}(x).$$

Let us explain the definition of the function f. Given any $x \in X_0$, there is an $i_x \in I$ such that $x \in X_{i_x}$. We may thus consider the value $f_{i_x}(x) \in Y_{i_x} \subseteq Y_0$. However, there may be another index $i'_x \in I$ with $x \in X_{i'_x}$, and a-priori,

$$f_{i'_x}(x) \neq f_{i_x}(x)$$

is a possibility. However, since $\{(X_i, f_i, Y_i)\}_{i \in I}$ is a chain, we may assume, without loss of generality, that $(X_{i_x}, f_{i_x}, Y_{i_x}) \leq (X_{i'_x}, f_{i'_x}, Y_{i'_x})$. But then, the definition of $\leq$ in P implies that

$$f_{i'_x}(x) = f_{i_x}(x)$$

and so the function $f : X_0 \to Y_0$ above is well-defined. A similar argument shows that f is bijective.

Now that we have verified the conditions of Zorn's Lemma we are guaranteed of the existence of a maximal element $(X_M, f_M, Y_M) \in P$. That is, $f_M : X_M \to Y_M$ is a bijective function for some $X_M \subseteq X$ and $Y_M \subseteq Y$. Let us now entertain the possibility that both X_M and Y_M are proper subsets. That is, that there exist elements $x_! \in X - X_M$ and $y_! \in Y - Y_M$. But then we may define

$$X_! = X_M \cup \{x_!\}$$
$$Y_! = Y_M \cup \{y_!\}$$

and $f_! : X_! \to Y_!$ by

$$f_!(x) = \begin{cases} f_M(x) & \text{if } x \in X_M \\ y_! & \text{if } x = x_! \end{cases}$$

giving rise to the element $(X_!, f_!, Y_!) \in P$ (the reader is invited to verify membership in P) with

$$(X_M, f_M, Y_M) < (X_!, f_!, Y_!),$$

contradicting the maximality of (X_M, f_M, Y_M).

We thus conclude that either $X_M = X$ or $Y_M = Y$. If $X_M = X$, then the composition $X \xrightarrow{f_M} Y_M \xrightarrow{\text{incl.}} Y$ with the inclusion function yields an injection $X \to Y$, thus showing that $|X| \leq |Y|$. If $Y_M = Y$, then the composition $Y \xrightarrow{f_M^{-1}} X_M \xrightarrow{\text{incl.}} X$ with the inclusion function is an injection $Y \to X$, showing that $|Y| \leq |X|$, and completing the proof. □

1.2.19 The Real Numbers

The set $\mathbb{R}$ of real numbers may be constructed in numerous different ways and can also be characterized axiomatically in different ways. We will not concern ourselves here with the construction of a model of the reals, and thus accept their existence and only state the axioms that govern them, namely that the reals form a Dedekind complete totally ordered field.

The statement that $\mathbb{R}$, with addition and multiplication, is a field is the claim that the following axioms hold:

- Associativity of addition—for all $a, b, c \in \mathbb{R}$: $(a + b) + c = a + (b + c)$.
- Commutativity of addition—for all $a, b \in \mathbb{R}$: $a + b = b + a$.
- Neutrality of 0—for all $a \in \mathbb{R}$: $a + 0 = a = 0 + a$.
- Existence of additive inverses—for all $a \in \mathbb{R}$ there exists an element $b \in \mathbb{R}$ with $a + b = 0 = b + a$.
- Associativity of multiplication—for all $a, b, c \in \mathbb{R}$: $(a \cdot b) \cdot c = a \cdot (b \cdot c)$.
- Commutativity of multiplication—for all $a, b \in \mathbb{R}$: $a \cdot b = b \cdot a$.
- Neutrality of 1—for all $a \in \mathbb{R}$: $1 \cdot a = a = a \cdot 1$.
- Existence of reciprocals—for all $a \in \mathbb{R}$, if $a \neq 0$, then there exists an element $b \in \mathbb{R}$ with $a \cdot b = 1 = b \cdot a$.
- Distributivity—for all $a, b, c \in \mathbb{R}$: $a \cdot (b + c) = a \cdot b + a \cdot c$.

In general, any set K with addition and multiplication operations, and distinguished elements $0 \neq 1$ in K, that satisfy these axioms is said to be a *field*. For instance, both $\mathbb{C}$ and $\mathbb{Q}$, with the usual notion of addition and multiplication, are fields. In contrast, $\mathbb{Z}$ with the same familiar addition and multiplication is not a field.

The statement that the reals form an ordered field is the claim that with the usual notion of $a \leq b$ for the real numbers, $\mathbb{R}$ is a totally ordered set, and that the following axioms hold:

- For all $a, b, c \in \mathbb{R}$: $a \leq b \implies a + c \leq b + c$.
- For all $a, b, c \in \mathbb{R}$ with $c \geq 0$: $a \leq b \implies a \cdot c \leq b \cdot c$.

In general, a field K with a total order on it such that these two axioms hold is called an *ordered field*. For instance, $\mathbb{Q}$ with the usual ordering is an ordered field. For the field $\mathbb{C}$, however, it can be shown without too much difficulty that no ordering of it exists that turns it into an ordered field.

Finally, perhaps the most important property of the real numbers, and certainly one that sets it apart among all ordered fields, is its Dedekind completeness. Recall that an upper bound for a subset $S \subseteq \mathbb{R}$ is a real number a such that $s \leq a$ for all $s \in S$. A *supremum* (or *least upper bound*) of S is an upper bound a with the property that if b is any upper bound of S, then $a \leq b$. Similarly, one defines the *infimum* (or *greatest lower bound*) of S to be a lower bound which is not smaller than any other lower bound. The statement that $\mathbb{R}$ is *Dedekind complete* is the claim: any non-empty bounded above subset $S \subseteq \mathbb{R}$ has a supremum. It can then be shown that this property is equivalent to: any non-empty bounded below subset $S \subseteq \mathbb{R}$ has an infimum.

In general, any Dedekind complete ordered field can be taken as essentially the field of the real numbers. In other words, the axioms listed above of a Dedekind complete ordered field determine the reals up to an isomorphism, meaning that the only difference between any two Dedekind complete ordered fields is the naming of the elements.

Further Reading

The closing section of this chapter (and of each of the forthcoming chapters) consists of suggestions for further reading. The list of sources is deliberately kept short and is thus by necessity not comprehensive. It aims to be a starting point for the reader interested in learning more about particular aspects of the chapter that were, for whatever reason, not elaborated upon in the main text.

For a broad historical perspective on the development of modern mathematics, including a detailed discussion of the birth of modern mathematical analysis, see [11]. The reader interested in the interplay between mathematics as a formal system and the real world, and the student baffled by the usefulness of mathematics in physics, is referred to [8]. For a discussion of the internal forces governing and influencing the development of mathematics see the classic text [7].

For a more in-depth treatment of set theory the reader may consult the book [9] which is also an introduction to logic and proof theory. This book also includes a detailed and elementary presentation of the Axiom of Choice and some of its equivalents, i.e., Zorn's Lemma, Zermelo's well-ordering principle, and the principle of cardinal comparability, as well as an elementary treatment of cardinal arithmetic. Various books (e.g., [6, 10]) address the Axiom of Choice from a historical point-of-view, discussing its influence on and within mathematics.

For more on the various possibilities of constructing the real numbers, see the review given in [5]. Some of these constructions refer to techniques of nonstandard analysis, i.e., to models of real numbers which contain actual infinitesimals. The reader interested in this somewhat controversial approach to analysis may consult the books [1, 3, 4] as well as [2] giving an eightfold path to the subject.

References

1. S. Albeverio, R. Høegh-Krohn, J.E. Fenstad, T. Lindstrøm, *Nonstandard Methods in Stochastic Analysis and Mathematical Physics*, vol. 122, Pure and Applied Mathematics (Academic Press, Orlando, 1986)
2. V. Benci, M. Forti, M. Di Nasso:, The eightfold path to nonstandard analysis. Nonstandard methods and applications in mathematics. Lecture Notes Log., Assoc. Symbol. Logic, vol. 25 (Jolla, Los Angeles, 2006), pp. 3–44
3. M. Davis, *Applied Nonstandard Analysis*, Pure and Applied Mathematics (Wiley, New York, 1977), p. 181
4. R. Goldblatt, *Lectures on the Hyperreals*, vol. 188, Graduate Texts in Mathematics (Springer, New York, 1998)
5. J. Harrison, *Constructing the Real Numbers in HOL, Higher order logic theorem proving and its applications*, IFIP Trans. A Comput. Sci. Tech., A-20 (Leuven, 1992), pp. 145–164
6. T.J. Jech, *The Axiom of Choice*, Dover Books on Mathematics (Dover, New York, 2008), p. 224
7. I. Lakatos, *Proofs and Refutations: The Logic of Mathematical Discovery* (Cambridge University Press, Cambridge, 1976)
8. S. Mac Lane, *Mathematics, Form and Function* (Springer, New York, 1986), p. 476
9. I. Moerdijk, J. van Oosten Sets, Models and Proofs, Lecture Notes, Utrecht University (2011), p. 109. http://www.staff.science.uu.nl/ooste110/syllabi/setsproofs09.pdf
10. G.H. Moore, *Zermelo's Axiom of Choice*, Dover Books on Mathematics (Dover, New York, 2013)
11. C. Smoryński, *Adventures in Formalism*, vol. 2, Texts in Mathematics (College Publications, London, 2012)

Chapter 2
Linear Spaces

Abstract This chapter is a rigorous introduction to linear spaces, but with a strong emphasis on infinite dimensional linear spaces. No prior knowledge of linear spaces is assumed, so that all definitions and proofs are included, but some mathematical maturity is assumed, dictating the level of detail given. In particular, all results (e.g., existence of dimension) are given in full generality using Zorn's Lemma.

Keywords Linear space · Vector space · Linear transformation · Operator · Dimension · Quotient linear space · Product linear space · Inner product space · Normed space · Cauchy-Schwarz inequality

This chapter assumes a rudimentary understanding of the linear structure of $\mathbb{R}^n$, a very rich structure, both algebraically and geometrically. Elements in $\mathbb{R}^n$, when thought of as vectors, that is as entities representing direction and magnitude, can be used to form parallelograms, can be scaled, the angle between two vectors can be computed, and the length of a vector can be found. These geometric features are given algebraically by means of, respectively, vector addition, scalar multiplication, the inner product of two vectors, and the norm of a vector. In this chapter these notions are abstracted to give rise to the concepts of linear space, inner product space, and normed space.

The chapter gives a detailed presentation of all of the relevant notions of linear spaces, provides examples, and contains rigorous proofs of all of the results therein. Exploiting the assumption of a rudimentary understanding of the linear structure of $\mathbb{R}^n$, and thus of finite dimensional linear spaces, the chapter has a clear secondary goal, namely to explore the subtleties of infinite dimensional linear spaces. This is an absolute necessity with Hilbert spaces in mind, since virtually all interesting examples of Hilbert spaces are infinite dimensional.

A consequence of this infinite dimensional theme of the chapter is that some proofs are considerably more involved than their finite dimensional counterparts. In particular, the theorems establishing the existence of bases and the concept of dimension are sophisticated and resort to an application of Zorn's Lemma. Also, cardinality considerations are important, since one needs to be able to compute, at least a little bit, with infinite quantities. To facilitate an easier reading of this chapter,

© Springer International Publishing Switzerland 2015
C. Alabiso and I. Weiss, *A Primer on Hilbert Space Theory*,
UNITEXT for Physics, DOI 10.1007/978-3-319-03713-4_2

the Preliminaries contain an account of Zorn's Lemma (Sect. 1.2.17) and some basic cardinal arithmetic (Sects. 1.2.12 and 1.2.13). Moreover, the text below indicates which proofs can safely be skipped on a first reading, leaving it to the discretion of the reader when, and whether, to tackle the technicalities involved in mastering the more advanced techniques.

Section 2.1 introduces the axioms of linear spaces in full generality, establishes basic properties and explores examples, most of which will be revisited throughout the book. Section 2.2 is concerned with establishing the notion of dimension for an arbitrary linear space. In particular, the dimension need not be finite, which, at times, necessitates some more intricate proofs. Section 2.3 discusses linear operators, the natural choice of structure preserving functions between linear spaces, studies their basic properties, and discusses the notion of isomorphic linear spaces. Section 2.4 introduces standard constructions producing new spaces from given ones, and in particular the kernel and image of a linear operator are discussed. The final section is devoted to inner product spaces and normed spaces and presents several important examples such as function spaces and ℓ_p spaces.

2.1 Linear Spaces—Elementary Properties and Examples

In $\mathbb{R}^2$, the scalar product $\alpha \cdot x$ is between a real number $\alpha \in \mathbb{R}$ and a vector $x \in \mathbb{R}^2$, and yields again a vector in $\mathbb{R}^2$. However, in similar situations, such as in $\mathbb{C}^2$, the scalar multiplication $\alpha \cdot x$ is now defined between any complex number $\alpha \in \mathbb{C}$ and an arbitrary vector x. The most general situation is when α is allowed to vary over the elements of an arbitrary (but fixed) *field* K. Prominent examples of fields are the field $\mathbb{R}$ of real numbers and the field $\mathbb{C}$ of complex numbers (for a more detailed discussion of fields the reader is referred to Sect. 1.2.19 of the Preliminaries). The definition of linear space given below is the result of a distillation of certain key properties of vector addition and scalar multiplication in $\mathbb{R}^2$ or $\mathbb{R}^3$, and is formalized in the most general form, namely with arbitrary fields. If the reader is not familiar with any fields other than $\mathbb{R}$ or $\mathbb{C}$ (or $\mathbb{Q}$, the field of rational numbers), then it is perfectly safe to proceed and replace any occurrence of an arbitrary field K by either $\mathbb{R}$ or $\mathbb{C}$. In the context of this book these are in any case the most important fields.

Definition 2.1 *(Linear Space)* Let K be an arbitrary field (such as the field of real numbers or of complex numbers, for concrete examples). A set V of elements $x, y, z, \ldots$ of an arbitrary nature, together with an operation, called *vector addition*, or simply *addition*, associating with any two elements $x, y \in V$ an element $z \in V$, called the *sum* of x and y, and denoted by $z = x + y$, as well as an operation associating with any $x \in V$ and $\alpha \in K$ an element $w \in V$, called the *product* or *scalar product* of α and x, and denoted by $w = \alpha \cdot x$, is called a *linear space* if

1. For all $x, y, z \in V$, the operation of vector addition satisfies:
   - Associativity, i.e., $(x + y) + z = x + (y + z)$.

- Commutativity, i.e., $x + y = y + x$.
- Existence of a *neutral* element, i.e., there exists an element $0 \in V$, for which

$$x + 0 = x = 0 + x.$$

- Existence of *additive inverses*, i.e., there exists an element $x' \in V$ such that

$$x + x' = 0 = x' + x.$$

2. For all $x \in V$ and $\alpha, \beta \in K$, the scalar product operation satisfies:
   - Associativity, i.e., $\alpha \cdot (\beta \cdot x) = (\alpha\beta) \cdot x$.
   - Neutrality of $1 \in K$, i.e., $1 \cdot x = x$.
3. For all $x, y \in V$ and $\alpha, \beta \in K$, the scalar product and vector addition operations are compatible in the sense that
   - Scalar product distributes over vector addition

$$\alpha \cdot (x + y) = \alpha \cdot x + \alpha \cdot y.$$

   - Scalar product distributes over scalar addition

$$(\alpha + \beta) \cdot x = \alpha \cdot x + \beta \cdot x.$$

A linear space is also called a *vector space*, its elements are called *vectors*, and the elements of the field K are called *scalars*. If we wish to emphasize the relevant field, then we say that V is a *linear space over* K. Otherwise, the assertion that V is a linear space includes the implicit introduction of a field K serving as the field of scalars for V.

Scalars will typically be denoted by lower-case Greek letters from the beginning of the alphabet, namely α, β, γ, and so on, while vectors will be denoted by x, y, z, etc. In either case, subscripts or superscripts may be used to enhance readability.

2.1.1 Elementary Properties of Linear Spaces

We now turn to establish several properties of linear spaces that immediately follow from the axioms.

Proposition 2.1 *In any linear space V the following statements hold.*

1. *The neutral element $0 \in V$ is unique.*
2. *For all $x \in V$, the additive inverse x' is unique.*
3. *For all $\alpha \in K$ and $x \in V$, the equation $\alpha \cdot x = 0$ holds if, and only if, $\alpha = 0$ or $x = 0$.*

Proof

1. Suppose that $0' \in V$ is a neutral element. That is

$$x + 0' = x$$

for all $x \in V$, and thus

$$0 = 0 + 0' = 0' + 0 = 0'.$$

2. Suppose that $x'' \in V$ satisfies that $x + x'' = 0$, then

$$x' = x' + 0 = x' + (x + x'') = (x' + x) + x'' = 0 + x'' = x''.$$

3. For $\alpha = 0$

$$\alpha \cdot x = 0 \cdot x = (0 + 0) \cdot x = 0 \cdot x + 0 \cdot x \implies 0 \cdot x = 0.$$

For $x = 0$

$$\alpha \cdot x = \alpha \cdot 0 = \alpha \cdot (0 + 0) = \alpha \cdot 0 + \alpha \cdot 0 \implies \alpha \cdot 0 = 0.$$

In the other direction, if $\alpha \cdot x = 0$ and $\alpha \neq 0$, then upon multiplication by α^{-1}, one obtains

$$x = 1 \cdot x = (\alpha^{-1} \cdot \alpha) \cdot x = \alpha^{-1} \cdot (\alpha \cdot x) = \alpha^{-1} \cdot 0 = 0. \qquad \square$$

Remark 2.1 It similarly follows that for any vector x, the additive inverse x' is given by $x' = (-1) \cdot x$. It is further common to neglect the $\cdot$ denoting the scalar product, write $x' = -x$ and resort to the familiar conventions for algebraic manipulations on vectors and scalars commonly used for addition and multiplication, e.g., we write $x - y$ for $x + (-y)$ or $x + y + z$ for $(x + y) + z$, and so on. This convention, of course, considerably shortens proofs such as those given above and will be silently used throughout the text below.

Remark 2.2 Most of the linear spaces we shall be concerned with will be over the field $\mathbb{R}$ of real numbers, in which case they are called *real linear spaces*, or over the field $\mathbb{C}$ of complex numbers, in which case they are called *complex linear spaces*. In the absence of further specification, or as implied by context, a linear space is assumed to be either a real or a complex linear space.

Remark 2.3 We make no notational distinction between the *zero vector*, i.e., the neutral element with respect to vector addition, and the element 0 in the field K. This is a generally safe practice, since context typically immediately points to the correct interpretation. For instance, in the equation $0 \cdot x = 0$, context dictates that the 0 on the left-hand-side is the scalar $0 \in K$ while on the right-hand-side it is the zero vector $0 \in V$.

2.1.2 Examples of Linear Spaces

Since linear spaces are very common in mathematics, presenting an exhaustive list of linear spaces is a daunting task. The chosen examples below are meant to present some commonly occurring linear spaces, to explore some less common possibilities, and to familiarize the reader with some linear spaces of great importance in the context of this book. The latter refers to linear spaces of sequences and of functions, linear spaces that will be revisited throughout the rest of the text.

Example 2.1 In the familiar spaces $\mathbb{R}^2$ and $\mathbb{R}^3$, which are the mathematical models of the physical plane and space, vector addition has the geometric interpretation known as the parallelogram law, while scalar multiplication αx has a scaling effect on the vector x determined by the magnitude and sign of α.

Example 2.2 Given a natural number $n \geq 1$, let $\mathbb{R}^n$ be the set of n-tuples of real numbers. Thus if $x \in \mathbb{R}^n$, then it is of the form $x = (x_1, \dots, x_n)$ where (for every $1 \leq k \leq n$) x_k, the k-th *component of* x, is a real number.

For all $x, y \in \mathbb{R}^n$ and $\alpha \in \mathbb{R}$ defining

$$
\begin{aligned}
x + y &= (x_1 + y_1, x_2 + y_2, \dots, x_n + y_n) \\
\alpha x &= (\alpha x_1, \alpha x_2, \dots, \alpha x_n),
\end{aligned}
$$

endows $\mathbb{R}^n$ with the structure of a linear space over the field $\mathbb{R}$, as is easy to verify. Obviously, the cases $n = 2$ and $n = 3$ recover the familiar linear spaces $\mathbb{R}^2$ and $\mathbb{R}^3$ (respectively). Analogously, the set $\mathbb{C}^n$ of all n-tuples of complex numbers, with similar coordinate-wise operations, is a linear space over the field $\mathbb{C}$.

Example 2.3 Let $\mathbb{R}^\infty$ be the set of infinite sequences of real numbers. Thus, a typical element $x \in \mathbb{R}^\infty$ is of the form $x = (x_1, x_2, \dots, x_k, \dots)$ where $x_k \in \mathbb{R}$, for each $k \geq 1$, is a real number, the k-th *component* of x. For all $x, y \in \mathbb{R}^\infty$ and $\alpha \in \mathbb{R}$ setting

$$
\begin{aligned}
x + y &= (x_1 + y_1, x_2 + y_2, \dots, x_k + y_k, \dots) \\
\alpha x &= (\alpha x_1, \alpha x_2, \dots, \alpha x_k, \dots)
\end{aligned}
$$

endows $\mathbb{R}^\infty$ with the structure of a linear space over the field $\mathbb{R}$, as is easily seen. Similarly, the set $\mathbb{C}^\infty$ of all infinite sequences of complex numbers, with similar coordinate-wise operations, is a linear space over the field $\mathbb{C}$.

Remark 2.4 The convention that for an element x in either $\mathbb{R}^n$, $\mathbb{C}^n$, $\mathbb{R}^\infty$, or $\mathbb{C}^\infty$ its k-th *component* is denoted by x_k (as illustrated in the preceding examples) will be used throughout this text. To refer to a sequence of such vectors we may thus use super scripts for the different vectors, e.g., $\{x^{(m)}\}_{m\geq 1}$, and then $x_k^{(m)}$ refers to the k-th component of the m-th vector.

Example 2.4 Consider the subset $c \subseteq \mathbb{C}^\infty$ consisting of all convergent sequences (here convergence is in the usual sense of convergence of sequences of complex numbers), let $c_0 \subseteq c$ be the subset consisting of all sequences that converge to 0, and let $c_{00} \subseteq c_0$ be the set

$$c_{00} = \{(x_1, \ldots, x_m, 0, 0, 0, \ldots) \mid m \geq 1, \quad x_1, \ldots, x_m \in \mathbb{C}\}$$

of all sequences that are eventually 0. With the same definition of addition and multiplication as in Example 2.3, each of these sets is easily seen to be a linear space over $\mathbb{C}$. Obviously, replacing $\mathbb{C}$ throughout by $\mathbb{R}$ yields similar linear spaces over $\mathbb{R}$.

Example 2.5 The constructions given above of the linear spaces $\mathbb{R}^n$, $\mathbb{C}^n$, $\mathbb{R}^\infty$, and $\mathbb{C}^\infty$ easily generalize to any field K and to any cardinality. Indeed, consider an arbitrary set B and an arbitrary field K. Recall from the Preliminaries (Sect. 1.2.8) that the set K^B is the set of all functions $x : B \to K$. For all such functions

$$x : B \to K, \quad y : B \to K$$

and

$$\alpha \in K$$

define the functions $x + y : B \to K$ and $\alpha x : B \to K$ by

$$(x + y)(b) = x(b) + y(b), \quad (\alpha x)(b) = \alpha \cdot x(b).$$

With these notions of addition and scalar multiplication, the set K^B is easily seen to be a linear space over K. In particular, we obtain the linear spaces $\mathbb{R}^{\{1,2,\ldots,n\}} = \mathbb{R}^n$, $\mathbb{C}^{\{1,2,\ldots,n\}} = \mathbb{C}^n$, $\mathbb{R}^\mathbb{N} = \mathbb{R}^\infty$, and $\mathbb{C}^\mathbb{N} = \mathbb{C}^\infty$. Further, restricting attention to the subset $(K^B)_0$ consisting only of those functions $x : B \to K$ for which $x(b) = 0$ for all but finitely many $b \in B$ also yields a linear space, with exactly the same definition for addition and scalar multiplication as in K^B. In particular the linear space c_{00} from Example 2.4 is recovered by noticing that $c_{00} = (\mathbb{C}^\mathbb{N})_0$. Note however that for a general field K no analogue of the linear spaces c or c_0 need exist since K need not have any notion of convergence.

Example 2.6 For $n \geq 0$, let P_n be the set of all polynomial functions, that is functions of the form $p(t) = a_n t^n + \cdots + a_1 t + a_0$, with real coefficients and degree at most n. With the ordinary operations

$$(p + q)(t) = p(t) + q(t), \quad (\alpha p)(t) = \alpha \cdot p(t)$$

taken as addition and scalar multiplication, it is immediate to verify that P_n is a linear space over $\mathbb{R}$. Removing the restriction on the degrees of the polynomials one obtains the set P of all polynomial functions with real coefficients which, again with the obvious notions of addition and scalar multiplication, forms a linear space over $\mathbb{R}$. One may also consider polynomials with coefficients in the field $\mathbb{C}$ of complex numbers, and obtain similar linear spaces over $\mathbb{C}$.

Example 2.7 Let I be a subset of $\mathbb{R}$ which is either an open interval (a, b), a closed interval $[a, b]$, or the entire real line $\mathbb{R}$. Consider the set $C(I, \mathbb{R})$ of all continuous real-valued functions $x : I \to \mathbb{R}$. The familiar definitions

$$(x + y)(t) = x(t) + y(t), \quad (\alpha x)(t) = \alpha \cdot x(t),$$

when applied to continuous functions $x, y : I \to \mathbb{R}$, are well-known to give continuous functions again, and it is easy to see that when these operations are taken as addition and scalar multiplication, the set $C(I, \mathbb{R})$ is a linear space over $\mathbb{R}$. One may also consider the set $C(I, \mathbb{C})$ of all continuous complex-valued functions $x : I \to \mathbb{C}$ to similarly obtain a linear space over $\mathbb{C}$. One may also consider, for each $k \geq 1$, the set $C^k(I, \mathbb{R})$ of all functions $x : I \to \mathbb{R}$ with a continuous k-th derivative, which is similarly a linear space over $\mathbb{R}$. It is then customary to equate $C^0(I, \mathbb{R})$ with $C(I, \mathbb{R})$. We may also allow $k = \infty$ so as to obtain $C^\infty(I, \mathbb{R})$, the linear space of all infinitely differentiable functions.

Example 2.8 Let K be any field and F a proper subfield of K, namely F is a proper subset of K, and with the induced operations from K it is itself a field. For instance, $\mathbb{Q}$ is a subfield of $\mathbb{R}$ which in turn is a subfield of $\mathbb{C}$. In such a situation any linear space V over the larger field K is also a *different* linear space over the smaller field F. The reason is that, of course, the additive structure of V is unaffected by the choice of field of scalars, whereas the axioms relating to the scalar product, if they hold for scalars ranging over K, then they certainly hold for scalars ranging over F (since all scalar axioms are universal equational quantifications). The process of considering a linear space over K as a linear space over F is named *restriction of scalars*.

In particular, each of the examples above of a linear space over $\mathbb{C}$ is also a linear space over $\mathbb{R}$, and also a linear space over $\mathbb{Q}$. Any linear space obtained by restriction of scalars $F \subset K$ is (except for trivial cases) very different than the original space. Another particular instance of restriction of scalars is the observation that since $\mathbb{R}$ is a linear space over itself (if this is confusing pause for a minute to realize that this is essentially the assertion that $\mathbb{R}^1$ is a linear space), restriction of scalars implies that $\mathbb{R}$ is also a linear space over $\mathbb{Q}$.

Exercises

Exercise 2.1 Let V be a linear space over K and $f : V \to X$ a bijection where X is a set with no a-priori extra structure. Let $f^{-1} : X \to V$ be the inverse function of f. Prove that the operations

$$x + y = f\left(f^{-1}(x) + f^{-1}(y)\right)$$

and

$$\alpha x = f\left(\alpha f^{-1}(x)\right),$$

defined for all $x, y \in X$ and $\alpha \in K$, turn X into a linear space over K.

Exercise 2.2 For any $\alpha \in \mathbb{C}$ let c_α be the set of all sequences $(x_1, x_2, \ldots) \in \mathbb{C}^\infty$ which converge to α. Prove that the linear structure of $\mathbb{C}^\infty$ restricts to a linear structure on c_α if, and only, if $\alpha = 0$.

Exercise 2.3 Let K be a field (such as $\mathbb{R}$ or $\mathbb{C}$ for familiarity) and let $M_{n,m}(K)$ be the set of $n \times m$ matrices with entries in the field K. Prove that with ordinary matrix addition and scalar product, the set $M_{n,m}(K)$ is a linear space over K.

Exercise 2.4 With the usual operations of addition and multiplication, is the set $(\mathbb{R} - \mathbb{Q}) \cup \{0\}$ a linear space over $\mathbb{Q}$?

Exercise 2.5 Let V be a linear space over K and fix some vector $w_0 \in V$. Define on the set V an addition operation by $x \oplus y = x + y - w_0$ and a scalar product operation by $\alpha \odot x = \alpha(x - w_0)$. Prove that with these operations V is a linear space over K whose zero vector is w_0.

Exercise 2.6 Prove that the set $\mathbb{R}$ of real numbers with addition given by $x \oplus y = xy$ and scalar multiplication given by $\alpha \odot y = y^\alpha$ is a linear space over the field $\mathbb{R}$.

Exercise 2.7 Let X be a set with one element. Prove that for any field K there is a unique choice of operations that turns X into a linear space over K.

Exercise 2.8 Prove that any linear space over $\mathbb{R}$ has either just a single vector or infinitely many.

Exercise 2.9 Let V be a linear space. Prove that $-x = (-1) \cdot x$ holds for all vectors $x \in V$.

Exercise 2.10 Let K be a field and $\mathscr{B}$ an arbitrary set. Consider the set $\langle \mathscr{B} \rangle_K$ of all formal expressions of the form

$$\sum_{b \in \mathscr{B}} \alpha_b \cdot b$$

where $\alpha_b \in K$ for each $b \in \mathscr{B}$ and at most finitely many of the α_b are non-zero. Verify that the obvious way to define addition and scalar multiplication on the set $\langle \mathscr{B} \rangle_K$ turns it into a linear space over K (known as the *free linear space generated by* $\mathscr{B}$).

2.2 The Dimension of a Linear Space

The familiar linear spaces $\mathbb{R}^n$ are all infinite sets (in fact, they all have the same cardinality, the cardinality c of the continuum). However, it is intuitively clear that $\mathbb{R}^3$ is considerably 'larger' than, say, $\mathbb{R}^2$. This fact is usually expressed in the claim that $\mathbb{R}^3$ has dimension 3, while $\mathbb{R}^2$ has dimension 2. The notion of dimension in general linear spaces is rather subtle, especially for the infinite dimensional ones. We now attend to investigate this situation in full generality, starting with the related concepts of linear independence and spanning sets.

2.2.1 Linear Independence, Spanning Sets, and Bases

By forming linear combinations of vectors from any given set of vectors in a linear space, one obtains a (potentially) large collection of vectors. The original set is said to be linearly independent if, intuitively, it does not allow for any redundancies when forming linear combinations while it is a spanning set if it is sufficiently large to generate any vector as a linear combination. The precise definitions follow.

Definition 2.2 Given any set $S \subseteq V$ of vectors in a linear space V (S may be finite or infinite), a *linear combination* of elements of S is any vector of the form

$$x = \sum_{k=1}^{m} \alpha_k x_k$$

with $x_1, \ldots, x_m$ vectors from S and $\alpha_1, \ldots, \alpha_m \in K$ arbitrary scalars. Equivalently,

$$x = \sum_{s \in S} \alpha_s \cdot s$$

with $\alpha_s = 0$ for all but finitely many $s \in S$. The two forms are essentially the same, differing only notationally. The *span* of S is then the set of all linear combinations of elements from S, and S is said to be a *spanning set* if its span is the entire linear space V. A spanning set S is a *minimal spanning set* if it is itself a spanning set but no proper subset of it is a spanning set. Further, S is said to be a set of *linearly independent* vectors if the only possibility of expressing the zero vector as a linear combination of elements from S is the trivial linear combination, i.e., where all the coefficients are 0. That is, S is linearly independent if whenever one has

$$0 = \sum_{k=1}^{m} \alpha_k x_k,$$

with $x_1, \ldots, x_m$ vectors in S, then necessarily $\alpha_k = 0$ for all $1 \leq k \leq m$. The set S is said to be a *maximal linearly independent* set if it is itself linearly independent but any set that properly contains it is not linearly independent. A set that is not linearly independent is also referred to as a *linearly dependent* set. Finally, a set that is both a spanning set and is linearly independent is said to be a *basis* of the linear space.

Remark 2.5 We speak of vectors $x_1, \ldots, x_m \in V$ as being either spanning or linearly independent, if the set $\{x_1, \ldots, x_m\}$ is spanning or linearly independent. Of course, we may also consider countably infinitely many vectors $x_1, x_2, \ldots$ as being spanning or linearly independent, in a similar fashion.

Example 2.9 The situation in $\mathbb{R}^n$ is probably very familiar to the reader. Any m vectors $x_1, \ldots, x_m$ in $\mathbb{R}^n$ are linearly independent if, and only if, the equation

$$\sum_{k=1}^{m} \alpha_k x_k = 0$$

admits the unique solution $\alpha_1 = \alpha_2 = \cdots = \alpha_m = 0$. It is well-known that linear independence implies $m \leq n$. Similarly, the given vectors are spanning if, and only if, for every vector $b \in \mathbb{R}^n$ the equation

$$\sum_{k=1}^{m} \alpha_k x_k = b$$

admits a solution. It is again a familiar fact that if the given vectors are spanning, then $m \geq n$. It thus follows that a basis for $\mathbb{R}^n$ must consist of precisely n vectors. In particular, all bases have the same size, namely n, which is referred to as the dimension of $\mathbb{R}^n$. Below we prove that every linear space has a dimension, if we allow infinite cardinalities into the picture. The result in that generality subsumes the properties of $\mathbb{R}^n$ just mentioned.

Example 2.10 In the space $\mathbb{C}^2$, considered as a linear space over $\mathbb{C}$, the vectors $(1, 0)$ and $(0, 1)$ are immediately seen to form a basis. However, if $\mathbb{C}^2$ is considered as a linear space over $\mathbb{R}$ (by the procedure of restriction of scalars from Example 2.8), then these two vectors are (of course) still linearly independent but they fail to span $\mathbb{C}^2$. Indeed, since only real scalars may now be used to form linear combinations of these vectors, the span will only be $\mathbb{R}^2$. To obtain a basis, the two vectors need to be augmented, for instance, by the vectors $(i, 0)$ and $(0, i)$. The four vectors together do form a basis of $\mathbb{C}^2$.

Example 2.11 In the linear space $\mathbb{R}^n$ or $\mathbb{C}^n$, the vectors $x_1, \ldots, x_n$ where

$$x_k = (0, \ldots, 0, 1, 0, \ldots, 0)$$

with 1 in the k-th position, are easily seen to be spanning and linearly independent, and thus form a basis, called the *standard basis* of $\mathbb{R}^n$, respectively $\mathbb{C}^n$. It is obvious that $\mathbb{R}^n$ and $\mathbb{C}^n$ have infinitely many bases.

In the examples presented so far it was quite straightforward to obtain a basis. The following example shows that this is not always the case. In fact, it is not even clear that the next linear space even has a basis.

Example 2.12 Let $\mathbb{C}^\infty$ be the linear space from Example 2.3. The vectors $x_1, x_2, \ldots$, given by

$$x_k = (0, \ldots, 0, 1, 0, \ldots)$$

with 1 in the k-th position are easily seen to be linearly independent, but they do not form a spanning set. Indeed, since the span consists only of the *finite* linear combinations of vectors from the set, the span in this case is the set of all vectors of the form $(a_1, \ldots, a_k, 0, \ldots, 0, 0, \ldots)$, namely those infinite sequences of complex numbers that are eventually 0. In other words, the span in $\mathbb{C}^\infty$ of the vectors $x_1, x_2, \ldots$ is the space c_{00} from Example 2.4, and thus we incidentally found a basis for c_{00}. It is now tempting to proceed as follows. Taking any vector $y_1 \in \mathbb{R}^\infty$ which is not spanned by $x_1, x_2, \ldots$, for instance the vector $y_1 = (1, 1, 1, \ldots, 1, \ldots)$, forming the set $\{y_1, x_1, x_2, \ldots\}$ must get us closer to obtaining a basis. Indeed, the new set is still linearly independent precisely because y_1 was not spanned by the rest of the vectors. But, this new set is still not a basis as there are still many vectors it fails to span, for instance the vector $y_2 = (1, 0, 1, 0, 1, 0, \ldots)$. Of course, we may now consider the larger set $\{y_1, y_2, x_1, x_2, \ldots\}$, but it too fails to be a basis. One may attempt to resolve the argument once and for all by claiming that proceeding in this way to infinity will eventually result in a basis. However, this is a very vague statement, and even if this process can be carried out mathematically (which it can, as will be shown below), it is entirely unclear as to which vectors will end up in the basis and which will not. In other words, even if this linear space has a basis, it is unlikely we can ever present one.

Naturally, similar observations hold true for $\mathbb{R}^\infty$ instead of $\mathbb{C}^\infty$, and in fact to most of the linear spaces in this book, and in analysis in general.

Example 2.13 Recall (see Example 2.8) that one can consider the space $\mathbb{R}$ as a linear space over the field $\mathbb{Q}$ of rational numbers (as a particular case of restriction of scalars). A real number α is said to be *transcendental* if it is not the root of a polynomial with rational coefficients. Examples of transcendental numbers include e and π, though the proofs are far from trivial. For any transcendental number x the set $\{1, x, x^2, x^3, \ldots\}$ is linearly independent (this is basically the definition of x being transcendental), but it is not a spanning set.

Example 2.14 Recall the space P_n from Example 2.6 of all polynomial functions with real coefficients of degree at most n. For every $k \geq 0$ let p_k be the vector $p_k(t) = t^k$. Clearly, the vectors $p_0, p_1, p_2, \ldots, p_n$ form a basis for P_n, but again there are infinitely many other choices of bases. The space P of all polynomial functions with real coefficients has a countable basis given by $p_0, p_1, p_2, \ldots$, a fact that is easily verified. Noticing that polynomials are continuous functions, no matter on which interval they are defined, we see by the above that in the linear space $C(I, \mathbb{R})$ of all continuous functions $x : I \to \mathbb{R}$, where I is a non-degenerate interval (i.e., not reducing to a point) the vectors $p_0, p_1, p_2, \ldots$ are linearly independent. However, they do not form a basis since any linear combination of these vectors is again a polynomial function, but not all continuous functions are polynomial functions. Once more, it is not at all clear that $C(I, \mathbb{R})$ even has a basis (what would one look like?).

From the discussion above we see that in some linear spaces, such as $\mathbb{R}^n$, P_n, or P, it is quite easy to find bases while in other linear spaces, such as $\mathbb{R}^\infty$, $\mathbb{R}$ over $\mathbb{Q}$, or $C(I, \mathbb{R})$, it is a highly non-trivial task. In more detail, we saw that it is rather simple to exhibit a large set of linearly independent vectors, but it is not so easy to have these vectors also span the entire space (the spaces c_{00} and P, as discussed above, are the exception to the rule). In fact, it is not at all clear that one can find bases for V in the case of $\mathbb{R}^\infty$ or $\mathbb{C}^\infty$, as well as for $\mathbb{R}$ as a linear space over $\mathbb{Q}$, or for $C(I, \mathbb{R})$.

2.2.2 Existence of Bases

In order to establish that every linear space does have a basis, we observe some immediate facts. It should be noted at once that the existence proof uses Zorn's Lemma in an essential way. That is, it can be shown that if every linear space has a basis, then the Axiom of Choice holds. Consequently, the existence proof is not constructive.

Proposition 2.2 *Let $S \subseteq V$ be a set of vectors in a linear space V. The following conditions are equivalent.*

1. *S is a maximal linearly independent set.*
2. *S is a minimal spanning set.*
3. *S is a basis.*

Proof First we show that if S is a maximal linearly independent set, then it is a basis. All that is needed is to show that S is a spanning set. To that end, let $x \in V$ be a vector in the ambient linear space. If $x \in S$, then it is certainly spanned by S. If $x \notin S$, then by virtue of S being a maximal linearly independent set, the set $S \cup \{x\}$ is linearly dependent. Thus, there exist vectors $x_1, \ldots, x_m \in S \cup \{x\}$ and non-zero scalars $\alpha_1, \ldots, \alpha_m \in K$ with

$$0 = \sum_{k=1}^{m} \alpha_k x_k.$$

However, in the expression above it must be that x itself appears as a summand since otherwise we would have expressed 0 as a non-trivial linear combination of vectors from the linearly independent set S. We may thus isolate x in the expression above to obtain it as a linear combination of elements from S. As x was arbitrary, we conclude that S is a spanning set.

Next we show that if S is a minimal spanning set, then it is a basis. All that is needed is to show that S is linearly independent, and indeed, if it were linearly dependent, then we would have some expression as above, giving 0 as a non-trivial linear combination of vectors from S. Using that expression we are able to isolate some vector $x \in S$ and exhibit it as a linear combination of other vectors from S. It is then easy to see that $S - \{x\}$ is still a spanning set (simply since any linear combination containing x can be replaced by one that does not). But this contradicts S being a minimal spanning set, and thus S must be linearly independent.

So far we have shown that each of conditions 1 and 2 implies condition 3. The proof will be completed by showing the converse of these implications. The details are very similar in spirit to those given so far, and thus the rest of the proof is left for the reader. □

We are now ready to establish that every linear space has a basis. The proof makes essential use of Zorn's Lemma (see Sect. 1.2.17 of the Preliminaries) and the reader may safely choose to skip the proof on a first reading. To the reader interested in the technique of Zorn's Lemma we remark that the proof below is actually a straightforward application with little technical difficulties, and is thus a fortunate first encounter with this important proof technique.

Theorem 2.1 *Every linearly independent set $A \subseteq V$ in a linear space V can be extended to a maximal linearly independent set. In particular, by considering the linearly independent set $\emptyset \subseteq V$ and by Proposition* 2.2, *it follows that every linear space V has a basis.*

Proof Consider the set P of all linearly independent subsets S of V that contain A, which we order by set inclusion. Evidently P is a poset, and to find a maximal linearly independent set that extends A amounts to finding a maximal element in P, and so we apply Zorn's Lemma (Sect. 1.2.17 of the Preliminaries). First we note that P is certainly not empty since clearly $A \in P$. Now, assume that $\{S_i\}_{i \in I}$ is a chain in P. We will show that $S = \bigcup_{i \in I} S_i$ is an upper bound for the chain. Clearly $S_i \subseteq S$ for all $i \in I$, thus the only thing to show is that $S \in P$, namely that S contains A (which is immediate) and that it is linearly independent. For that, assume that

$$0 = \sum_{k=1}^{m} \alpha_k x_k$$

is a non-trivial linear combination of the vectors $x_1, \ldots, x_m \in S$. Since S is the union of the S_i, it follows that $x_m \in S_{f(m)}$ for a suitable $f(m) \in I$, but since $\{S_i\}_{i \in I}$ is a chain it follows that there is a single index $i_0 \in I$ such that $x_1, \ldots, x_m \in S_{i_0}$. But then the equality above expresses the zero vector as a non-trivial linear combination of vectors from S_{i_0}, contradicting the fact that S_{i_0} is linearly independent. With that the conditions of Zorn's Lemma are satisfied, and so the existence of a maximal element in P is guaranteed. This maximal element is a set $S_M \in P$, namely S_M contains A and S_M is a maximal set of linearly independent vectors, as required. □

2.2.3 Existence of Dimension

Now that we know that every linear space has at least one basis, it is tempting to define the dimension of a linear space to be the cardinality of its basis. However, for this to make sense we need to know that all bases have the same cardinality. While this is a very plausible assertion (certainly for such familiar spaces as $\mathbb{R}^n$), it requires a careful proof, particularly in the infinite dimensional case. The important ingredient is the following lemma, which states that the cardinality of any linearly independent set is not greater than the cardinality of any spanning set. This result again uses Zorn's Lemma and its proof may safely be skipped on a first reading. A word of caution to the ambitious reader that this proof, compared to the proof of Theorem 2.1, is technically more demanding.

Lemma 2.1 *If V is a linear space, $I \subseteq V$ a linearly independent set of vectors, and $S \subseteq V$ a spanning set, then there exists an injective function $f : I \to S$.*

Proof Consider a pair (J, f) where $J \subseteq I$ and $f : J \to S$ is an injection. The idea will be (using Zorn's Lemma) to keep on extending the domain of f until the entire set I is exhausted. An important ingredient for achieving that is the following. Thinking of the injective function $f : J \to S$ as an instruction to replace the vectors in J by their images in S, we only consider such pairs (J, f) for which $(I - J) \cup f(J)$ is still a linearly independent set. Let us now form the set P of all such pairs and introduce an ordering on it declaring that $(J_1, f_1) \leq (J_2, f_2)$ precisely when $J_1 \subseteq J_2$ and f_2 extends f_1 (the latter means that $f_1(x) = f_2(x)$ holds for all $x \in J_1$). It is immediate that P is a poset, and that a maximal element in it will furnish us with an injective function $f : J_M \to S$ for a very large subset $J_M \subseteq I$. We will then show that necessarily $J_M = I$, and the result will be established.

To verify the conditions of Zorn's Lemma, notice first that P is not empty. Indeed, the pair $(\emptyset, \emptyset)$ is in P. Next, let $\{(J_t, f_t)\}_{t \in T}$ be a chain in P, for which we must now find an upper bound. Let

$$J = \bigcup_{t \in T} J_t$$

and notice that we may define $f : J \to S$ as follows. Given $x \in J$ there is some $t \in T$ such that $x \in J_t$, and so let $f(x) = f_t(x)$. To see that this is a well-defined function, i.e., that it is independent of the choice of $t \in T$, note that if $x \in J_{t'}$, then either $J_t \subseteq J_{t'}$ or $J_{t'} \subseteq J_t$ and then either $f_{t'}$ extends f_t or f_t extends $f_{t'}$, and in either case $f_t(x) = f_{t'}(x)$. It is clear that $J_t \subseteq J$ and that f extends f_t, for all $t \in T$, so all we need to do in order to show that (J, f) is an upper bound for the chain is establish that $(J, f) \in P$. Clearly, $J \subseteq I$, and $f : J \to S$ is injective (since we saw that in fact for any two $x_1, x_2 \in J$ the function f agrees with some f_t, which is injective, so $f(x_1) = f(x_2) \implies f_t(x_1) = f_t(x_2) \implies x_1 = x_2$). So it remains to observe that $(I - J) \cup f(J)$ is a linearly independent set of vectors. Indeed, if 0 can be obtained as a non-trivial linear combination of the vectors $x_1, \ldots, x_m \in (I - J) \cup f(J)$, then, using the chain condition again, it follows that there exists an element $t \in T$ such that $x_1, \ldots, x_m \in (I - J_t) \cup f_t(J_t)$, which is a linearly independent set, yielding a contradiction.

With the conditions of Zorn's Lemma now verified, it follows that there exists a maximal element (J_M, f_M), with $J_M \subseteq I$ and $f_M : J_M \to S$ an injection. If $J_M = I$, then we are done, since $f_M : I \to S$ is the required injection. Assuming this is not the case, let $x_! \in I - J_M$. If we can find a vector $y_! \in S - f_M(J_M)$ such that

$$(I - (J_M \cup \{x_!\})) \cup (f_M(J_M) \cup \{y_!\})$$

is linearly independent, then we will have that $(J_M \cup \{x_!\}, f_!) \in P$, where $f_!$ is the extension of f_M given by $f_!(x_!) = y_!$. But that would contradict the maximality of (J_M, f_M), and we will have our contradiction. So, we proceed to prove the existence of such a vector $y_!$. If no such $y_!$ exists, then that means that the set

$$(I - (J_M \cup \{x_!\})) \cup (f_M(J_M) \cup \{y\})$$

is a linearly dependent set for every $y \in S - f_M(J_M)$. Now, since S is a spanning set we may write

$$x_! = \sum_{s \in S} \alpha_s \cdot s = \sum_{s \in f_M(J_M)} \alpha_s \cdot s + \sum_{s \in S - f_M(J_M)} \alpha_s \cdot s = x_1 + x_2$$

where the sum is a finite sum, so that $\alpha_s = 0$ for all but finitely many s, and we simply split the sum according to whether or not $s \in f_M(J_M)$. By our assumption, the set

$$(I - (J_M \cup \{x_!\})) \cup f_M(J_M)$$

is linearly dependent if any of the vectors $s \in S - f_M(J_M)$ is added to it. Thus, this set is linearly dependent if any linear combination of such vectors, such as x_2, is added to it. Further, since x_1 is in the span of $f_M(J_M)$ it follows that the set above is linearly dependent if $x = x_1 + x_2$ is added to it. But the latter is the set

$$(I - (J_M \cup \{x_!\})) \cup f_M(J_M) \cup \{x_!\} = (I - J_M) \cup f_M(J_M)$$

which is linearly independent. The proof is now complete. □

We are now in position to establish the following important result.

Theorem 2.2 *Let V be a linear space. If B_1 and B_2 are any two bases for V, then they have the same cardinality.*

Proof By definition of basis, B_1 is linearly independent and B_2 is a spanning set. By Lemma 2.1, there is an injective function $f : B_1 \to B_2$. By the same argument there is also an injective function $g : B_2 \to B_1$. It now follows from the Cantor-Schröder-Bernstein theorem (Theorem 1.1 in the Preliminaries) that the cardinalities of B_1 and B_2 are equal. □

Definition 2.3 The *dimension* of a linear space V, indicated by $\dim(V)$, is equal to the cardinality of a basis for it. Notice that by Theorem 2.1, at least one basis exists, and by Theorem 2.2 all bases have the same cardinality. Thus, the notion of dimension is well-defined. The linear space V is said to be *finite dimensional* if its dimension is finite, and *infinite dimensional* otherwise. A basis for an infinite dimensional linear space is also referred to as a *Hamel basis*.

Example 2.15 Finite dimensional linear spaces, by Examples 2.11 and 2.14, include the spaces $\mathbb{R}^n$ and $\mathbb{C}^n$, having dimension n, and the space P_n of polynomials (see Example 2.6), which is of dimension $n+1$. Infinite dimensional linear spaces include, due to Examples 2.12 and 2.14, the space c_{00} of sequences that are eventually 0, the space P of all polynomials with real coefficients, and the space $C(I, \mathbb{R})$ of continuous function $x : I \to \mathbb{R}$, as long as I does not reduce to a point.

We have already encountered linear spaces of very large dimension. To argue about the dimension of such spaces (and in general) the following result, interesting on its own, is useful. Recall from Example 2.5 the linear space $(K^B)_0$ of all functions $x : B \to K$ satisfying $x(b) = 0$ for all but finitely many $b \in B$.

Proposition 2.3 *Let V be a linear space and B a basis for it. Then every vector $x \in V$ can be expressed uniquely as a linear combination of elements from B. In other words, there is a bijective correspondence between the vectors $x \in V$ and the elements of $(K^B)_0$.*

Proof Since a typical element in $(K^B)_0$ is nothing but a function $x : B \to K$ for which $x(b) = 0$ for all but finitely many $b \in B$, we may associate with each such x the vector $\sum_{b \in B} x(b) \cdot b$ (as the summation is finite). We claim that this correspondence is the desired bijection. Indeed, it is a tautology that the surjectivity of this process is precisely the claim that B is a spanning set, while it is almost a tautology that the injectivity of the process is the claim that B is linearly independent. □

We thus see that a basis B endows a linear space with a notion of coordinates. This is certainly a useful thing to have, but for practical reasons it is only as useful as the ability to explicitly describe the basis B. In finite dimensional linear spaces it is very common to work with bases but, as we remarked earlier, in infinite dimensional linear spaces being able to explicitly describe a basis is the exception rather than the rule. Consequently, Hamel bases in infinite dimensional linear spaces are generally used for theoretical rather than practical purposes.

Example 2.16 Recall from Example 2.8 that $\mathbb{R}$ may be viewed as a linear space over $\mathbb{Q}$. Suppose that $\mathscr{B}$ is a Hamel basis for $\mathbb{R}$ over $\mathbb{Q}$. By Proposition 2.3, there is then a bijection between $\mathbb{R}$ and the set $(\mathbb{Q}^{\mathscr{B}})_0$ of all functions $\mathscr{B} \to \mathbb{Q}$ which attain 0 at all but finitely many arguments. In other words, $|\mathbb{R}| = |(\mathbb{Q}^{\mathscr{B}})_0|$ (see Sect. 1.2.11 of the Preliminaries for the basics of cardinalities). If $\mathscr{B}$ is countable, then it is easy to write $(\mathbb{Q}^{\mathscr{B}})_0$ as a countable union of countable sets (refer to Sect. 1.2.13 from the Preliminaries for the relevant material) and thus the set $(\mathbb{Q}^{\mathscr{B}})_0$ itself would be countable. But that would imply that $\mathbb{R}$ is a countable set while the reals are well-known to be uncountable. We conclude that $\mathbb{R}$, as a linear space over $\mathbb{Q}$, is infinite dimensional of uncountable dimension.

We close this section by illustrating a difference between finite dimensional linear spaces and infinite dimensional ones.

Proposition 2.4 *The cardinality of any linearly independent set A in a linear space V is a lower bound for the dimension of V. Moreover, if V is finite dimensional, then any linearly independent set $A \subseteq V$ whose cardinality is equal to the dimension of V is a basis.*

Proof The dimension of V is the cardinality of any basis B, and as B is in particular a spanning set, it follows from Lemma 2.1 that there is an injection $A \to B$, and so the cardinalities satisfy $|A| \leq |B|$ (refer to Sect. 1.2.11 of the Preliminaries, if needed). The claim now follows since the latter is the dimension of V.

Now, if V is finite dimensional, say of dimension n, and $A = \{x_1, \ldots, x_n\}$ is a set of n linearly independent vectors, then to show A is a basis we just need to prove that it is a spanning set. But if it were not, and $y \in V$ is any vector not in its span, then the set $\{x_1, \ldots, x_n, y\}$ is linearly independent and contains $n+1$ vectors. But then, by the first part of the proposition, it would follow that $n + 1 \leq n$, an absurdity.

□

Remark 2.6 The finite dimensionality assumption is crucial. For instance, in the linear space c_{00} of sequences which are eventually 0 (Example 2.4), consider the vectors $\{x_k\}_{k \geq 1}$ where $x_k = (0, \ldots, 0, 1, 0, \ldots)$, with 1 in the k-th position. These vectors are easily seen to be linearly independent and spanning, thus they form a basis of countably many vectors. The dimension of c_{00} is thus infinitely countable. The set $\{x_2, x_3, \ldots\}$ is clearly also linearly independent, has the same cardinality as the dimension of c_{00}, but it is not spanning, and thus not a basis.

Exercises

Exercise 2.11 Let V be a linear space and $S, S' \subseteq V$ two arbitrary sets of vectors with $S \subseteq S'$. Prove that if S' is linearly independent, then so is S, and prove that if S is a spanning set, then so is S'.

Exercise 2.12 Let V be a finite dimensional linear space and S a spanning set. Prove that S can be sifted to give a basis, that is, show that there exists a subset $S' \subseteq S$ such that S' is a basis for V.

Exercise 2.13 Let V be a linear space (not necessarily finite dimensional) and S a spanning set. Prove that S can be sifted to give a basis, that is, show that there exists a subset $S' \subseteq S$ such that S' is a basis for V.

Exercise 2.14 Consider $\mathbb{R}$ as a linear space over $\mathbb{Q}$ and let α be a transcendental number. Prove that the set $\{1, \alpha, \alpha^2, \alpha^3, \ldots\}$ is linearly independent. Prove that it is not a spanning set by showing that its span is a countable set.

Exercise 2.15 Consider $\mathbb{R}$ as a linear space over $\mathbb{Q}$. Prove that the dimension of $\mathbb{R}$ over $\mathbb{Q}$ is $|\mathbb{R}|$, the cardinality of the real numbers.

Exercise 2.16 Let V be a linear space and $S \subseteq V$ a set of vectors. For a scalar $\alpha \in K$, let us write

$$\alpha S = \{\alpha x \mid x \in S\}.$$

Assuming that $\alpha \neq 0$ is fixed, prove that S is linearly independent (respectively spanning, a basis) if, and only if, αS is linearly independent (respectively spanning, a basis).

Exercise 2.17 Let V be a linear space and $\{x_k\}_{k \in \mathbb{N}}$ countably many vectors in V. For all $m \in \mathbb{N}$, let

$$y_m = \sum_{k=1}^{m} x_k.$$

Prove that $\{x_k\}_{k \in \mathbb{N}}$ is linearly independent (respectively spanning, a basis) if, and only if, $\{y_k\}_{k \in \mathbb{N}}$ is linearly independent (respectively spanning, a basis).

Exercise 2.18 Let B be an arbitrary set and K a field. Consider the linear spaces K^B and $(K^B)_0$. For every $b_0 \in B$ let $x_{b_0} : B \to K$ be the function

$$x_{b_0}(b) = \begin{cases} 1 & \text{if } b = b_0, \\ 0 & \text{otherwise.} \end{cases}$$

Prove that the set $\{x_{b_0}\}_{b_0 \in B}$ is linearly independent in K^B. Is it a basis for K^B? Is it a basis for $(K^B)_0$?

Exercise 2.19 In the linear space $C(\mathbb{R}, \mathbb{R})$, find three vectors $x, y, z : \mathbb{R} \to \mathbb{R}$ such that $\{x, y, z\}$ is a linearly independent set while $\{x^2, y^2, z^2\}$ is linearly dependent. (Here x^2 refers to the function given by $x^2(t) = (x(t))^2$, and similarly for the other functions.)

Exercise 2.20 Consider the space c_0 of sequences of complex (or real, if you like) numbers that converge to 0. How many linearly independent vectors can you find in c_0?

2.3 Linear Operators

The definition and properties of the structure preserving functions between linear spaces form the topic of this section. The definitions (including that of a linear isomorphism) and the results are discussed and exemplified in the context of the linear spaces introduced above.

Definition 2.4 Let V and W be linear spaces over the same field K. A function $T : V \to W$ is said to be *additive* if

$$T(x + y) = T(x) + T(y)$$

for all vectors $x, y \in V$, and it is said to be *homogenous* if

$$T(\alpha x) = \alpha T(x)$$

for all vectors $x \in V$ and all scalars $\alpha \in K$. If T is both additive and homogeneous, then it is called a *linear operator*, a condition equivalent to the equality

$$T(\alpha_1 x_1 + \alpha_2 x_2) = \alpha_1 T(x_1) + \alpha_2 T(x_2)$$

for all $x_1, x_2 \in V$ and $\alpha_1, \alpha_2 \in K$.

Remark 2.7 Synonymous terms for linear operator are *linear transformation* and *linear homomorphism*. Throughout the book we will adopt the convention that any reference to a linear operator $T : V \to W$ immediately implies, implicitly at times, that V and W are linear spaces over the same field K.

It is a straightforward proof by induction that a linear operator T preserves any linear combination, that is

$$T(\sum_{k=1}^{m} \alpha_k x_k) = \sum_{k=1}^{m} \alpha_k T(x_k)$$

for all scalars $\alpha_1, \ldots, \alpha_m \in K$ and vectors $x_1, \ldots, x_m \in V$. In fact, this property can equivalently be taken as the definition of linear operator. The next result shows that the linearity requirement of a linear operator forces any linear operator to also respect the zero vector, additive inverses, and subtraction.

Proposition 2.5 *For a linear operator $T : V \to W$:*

1. $T(0) = 0$.
2. $T(-x) = -T(x)$ *for all* $x \in V$.
3. $T(x - y) = T(x) - T(y)$ *for all* $x, y \in V$.

Proof

1. Notice that $T(0) + T(0) = T(0 + 0) = T(0)$, and now subtract $T(0)$ from both sides of the equation.
2. $T(x) + T(-x) = T(x + (-x)) = T(0) = 0$.
3. $T(x - y) = T(x + (-y)) = T(x) + T(-y) = T(x) - T(y)$. □

2.3.1 Examples of Linear Operators

Trivial, but important, examples of linear operators are the following. For any linear space V, the identity function $\mathrm{id} : V \to V$ is always a linear operator. On the other extreme, given linear spaces V and W over the same field, the function $T_0 : V \to W$ given by $T_0(x) = 0$ is also immediately seen to be a linear operator.

In the presence of a basis for the domain V, linear operators $T : V \to W$ are easily characterized by the following result, readily yielding an endless supply of examples.

Lemma 2.2 *Let V and W be linear spaces over the same field K, and let B be a basis for V. It then holds that any function $F : B \to W$ extends uniquely to a linear operator $T_F : V \to W$.*

Proof For a given function $F : B \to W$, suppose that $T_F : V \to W$ is a linear operator extending F. Then, given any vector $x \in V$, write

$$x = \sum_{b \in B} \alpha_b \cdot b$$

as a (finite) linear combination of basis elements, and then it follows that

$$T_F(x) = T_F(\sum_{b \in B} \alpha_b \cdot b) = \sum_{b \in B} \alpha_b \cdot T_F(b) = \sum_{b \in B} \alpha_b \cdot F(b).$$

Noticing that the computation above expresses $T(x)$ in terms of vectors of the form $F(b)$ we conclude that an extension T_F, if it exists, is unique. If we now take the equality above as a definition, then we obtain a function $T_F : V \to W$ (relying on the uniqueness of the linear combination expressing x in order to assure that T_F is well-defined). Verifying that this T_F is indeed a linear operator and that it extends F follow immediately. □

Remark 2.8 The linear operator $T_F : V \to W$ constructed from the given function $F : B \to W$ above is said to be obtained by *linearly extending* F to all of V.

Often enough the technique of the last result is inadequate, either because a basis is not readily available or because a more direct formula for the linear operator is obtainable. The following examples illustrate these possibilities.

Example 2.17 The derivation operation d/dt, taking a function f to its derivative df/dt, satisfies the well-known properties

$$\frac{d}{dt}(f+g) = \frac{d}{dt}(f) + \frac{d}{dt}(g)$$

and

$$\frac{d}{dt}(\alpha f) = \alpha \frac{d}{dt}(f),$$

namely, it is additive and homogenous, and thus, if we choose the domain and codomain correctly, we expect it to be a linear operator. To turn this observation into a precise statement, recall the linear spaces $C^k([a,b],\mathbb{R})$ from Example 2.7 of all functions $x : [a,b] \to \mathbb{R}$ with a continuous k-th derivative. Then, for every $k \geq 1$, the observation above can be stated by saying that $d/dt : C^k([a,b],\mathbb{R}) \to C^{k-1}([a,b],\mathbb{R})$ is a linear operator. Similarly, $d/dt : C^\infty([a,b],\mathbb{R}) \to C^\infty([a,b],\mathbb{R})$ is a linear operator on the linear space of infinitely differentiable functions.

Example 2.18 The integral operator $\int_a^b dt\, f(t)$ is also well known to be additive and homogenous, and, since every continuous function on a closed interval is integrable, we obtain that for every interval $I = [a,b]$ the function

$$f \mapsto \int_a^b dt\, f(t)$$

is a linear operator from the linear space $C(I,\mathbb{R})$ of Example 2.7 to $\mathbb{R}$ as a linear space over itself.

Example 2.19 Spaces of infinite sequences, such as $\mathbb{R}^\infty$ from Example 2.3 or the space c_0 from Example 2.4 admit the following operators (we use $\mathbb{R}^\infty$ just to fix one possibility). The *shift operator* $S : \mathbb{R}^\infty \to \mathbb{R}^\infty$, given, for $x = (x_1, x_2, \ldots, x_k \ldots)$, by $S(x) = (x_2, x_3, \ldots, x_k, \ldots)$, is easily seen to be a linear operator. It is equally easy to see that the function $T : \mathbb{R}^\infty \to \mathbb{R}^\infty$, given for $x = (x_1, x_2, \ldots, x_k \ldots)$ as above by $T(x) = (0, x_1, x_2, \ldots, x_k, \ldots)$ is a linear operator. Note that $S \circ T$ is the identity while $T \circ S$ is not, a phenomenon that is known to be impossible in finite dimensional linear spaces. These operators are the *creation* and *annihilation* operators $a^\dagger$ and a, widely used in Quantum Mechanics.

We see thus that even when a basis can be given explicitly (and certainly when it cannot) it may be much simpler to define a linear operator directly rather than by linear extension on a basis.

2.3.2 Algebra of Operators

With suitable domain and codomain, linear operators can be composed or added. We now investigate the resulting algebraic laws related to these operations.

Proposition 2.6 *The composition $S \circ T : U \to W$ of any two linear operators $U \xrightarrow{T} V \xrightarrow{S} W$ (where in particular all linear spaces are over the same field) is a linear operator.*

Proof The additivity of $S \circ T$ follows from the definition of the composition and the computation

$$S(T(x+y)) = S(T(x) + T(y)) = S(T(x)) + S(T(y)),$$

valid for all vectors $x, y \in U$, utilizing the additivity of T and S. The homogeneity of the composition follows similarly and the reader is invited to fill in the details. □

Next, any two functions $T, S : V \to W$ between linear spaces over the field K can be added to give rise to the function $T + S : V \to W$ given by

$$(T + S)(x) = T(x) + S(x)$$

for all $x \in V$. Further, given a scalar $\alpha \in K$, the function $\alpha T : V \to W$ is given by

$$(\alpha T)(x) = \alpha(T(x)).$$

Proposition 2.7 *For all linear operators $T, S : V \to W$ and scalars $\alpha \in K$, both $T + S$ and αT are linear operators.*

Proof For all vectors $x, y \in V$ and scalars $\beta, \gamma \in K$:

$$\begin{aligned}(T + S)(\beta x + \gamma y) &= T(\beta x + \gamma y) + S(\beta x + \gamma y)\\ &= T(\beta x) + T(\gamma y) + S(\beta x) + S(\gamma y)\\ &= \beta T(x) + \gamma T(y) + \beta S(x) + \gamma S(y)\\ &= \beta(T(x) + S(x)) + \gamma(T(y) + S(y))\\ &= \beta(T + S)(x) + \gamma(T + S)(y).\end{aligned}$$

The proof for αT is similar and left for the reader. □

The result above shows that the set of all linear operators $T : V \to W$ has naturally defined notions of addition and scalar multiplication. It is a pleasant fact that with these operations one obtains a linear space.

Definition 2.5 Given linear spaces V and W over the same field K, we denote by $\mathrm{Hom}(V, W)$ the set of all linear operators $T : V \to W$.

Theorem 2.3 *For linear spaces V and W over the same field K, the set* $\mathrm{Hom}(V, W)$, *when endowed with the operations of addition and scalar multiplication as above, is a linear space over K.*

Proof The proof is a straightforward verification of the linear space axioms, so we only give the details for a few of the axioms. For instance, given linear operators $T_1, T_2 : V \to W$, to show that

$$T_1 + T_2 = T_2 + T_1$$

we note that

$$(T_1 + T_2)(x) = T_1(x) + T_2(x) = T_2(x) + T_1(x) = (T_2 + T_1)(x)$$

where the commutativity of vector addition in W was used.

To show the existence of an additively neutral element, recall that $T_0 : V \to W$ given by $T_0(x) = 0$, is always a linear operator, and thus $T_0 \in \text{Hom}(V, W)$. Seeing that T_0 is additively neutral is simply the computation

$$(T + T_0)(x) = T(x) + T_0(x) = T(x) + 0 = T(x).$$

We leave the verification of the other axioms to the reader. □

2.3.3 Isomorphism

The linear spaces $\mathbb{R}^{n+1}$ and P_n, while consisting of very different elements, are, in a strong sense, essentially identical. It is obvious that one may think of a sequence of $n + 1$ real numbers as the coefficients of a polynomial, and, vice versa, one may identify a polynomial function with its list of coefficients and thus obtain $n + 1$ real numbers. Thus, one may rename the elements in one space to obtain the elements of the other, and, moreover, the linear structure under this renaming is respected. This situation is made precise, and generalized, by the concept of isomorphism.

Definition 2.6 A linear operator $T : V \to W$ which, as a function, is a bijection is called a *linear isomorphism* or (simply an *isomorphism* if the linear context is clear). When $T : V \to W$ is an isomorphism, the spaces V and W are said to be *isomorphic*, denoted by $V \cong W$.

The following result establishes some expected behaviour of isomorphisms.

Proposition 2.8 *Suppose that U, V, and W are linear spaces over the same field K and consider the linear operators $T : U \to V$ and $S : V \to W$. The following then hold.*

1. *The identity function $id : V \to V$ is an isomorphism.*
2. *If T and S are isomorphisms, then so is $S \circ T$.*
3. *If T is an isomorphism, then so is the inverse function T^{-1}.*

Proof

1. We already noted that the identity function is always a linear operator. It is clearly bijective and thus an isomorphism.
2. We already noted that the composition of linear operators is a linear operator, and thus $S \circ T$ is a linear operator. In general, the composition of bijective functions is again bijective, and so if T and S are isomorphisms, then so is $S \circ T$.
3. Since $T : U \to V$ is an isomorphism, the inverse function $T^{-1} : V \to U$ exists, and is clearly a bijection. To conclude the proof it remains to be shown that T^{-1} is a linear operator. Indeed, T^{-1} is additive since

$$\begin{aligned} T^{-1}(x+y) &= T^{-1}(T(T^{-1}(x)) + T(T^{-1}(y))) \\ &= T^{-1}(T(T^{-1}(x) + T^{-1}(y))) \\ &= T^{-1}(x) + T^{-1}(y) \end{aligned}$$

for all vectors $x, y \in V$. Further, T^{-1} is homogenous since

$$T^{-1}(\alpha x) = T^{-1}(\alpha T(T^{-1}(x))) = T^{-1}(T(\alpha T^{-1}(x))) = \alpha T^{-1}(x)$$

for all vectors $x \in V$ and scalars $\alpha \in K$. □

Corollary 2.1 *For all linear spaces U, V, and W over the same field K:*

1. $V \cong V$.
2. *If* $U \cong V$, *then* $V \cong U$.
3. *If* $U \cong V$ *and* $V \cong W$, *then* $U \cong W$.

Isomorphic linear spaces are essentially identical, except (possibly) for the names of the elements in them and they thus possess exactly the same linear properties. This is a somewhat vague statement but it is almost always immediate how to turn it into a precise statement. For instance, the dimension of a linear space is a linear property and thus any two isomorphic linear spaces have the same dimension, as we now show.

Proposition 2.9 *If $V \cong W$, then both linear spaces have the same dimension.*

Proof By assumption there exists a linear isomorphism $T : V \to W$. The dimension of a linear space is the cardinality of any of its bases, and thus the result will be established by exhibiting a bijection between a basis for V and a basis of W. Let $B \subseteq V$ be a basis and consider its image $T(B) \subseteq W$. Clearly, $T|_B$, the restriction of T to B, is a bijection between B and $f(B)$, so all that remains to be done is to show that $f(B)$ is a basis for W, namely that $f(B)$ is a spanning set of linearly independent vectors. Let $y \in W$ be an arbitrary vector. Since T is onto, we may write $y = T(x)$ for some $x \in V$. As B is a basis, we express x as a (finite!) linear combination of basis elements:

$$x = \sum_{b \in B} \alpha_b \cdot b.$$

But then

$$y = T(x) = \sum_{b \in B} \alpha_b \cdot T(b)$$

and thus y is in the span of $T(B)$. As y was arbitrary we conclude that $T(B)$ spans all of W. The proof that $T(B)$ is also linearly independent follows a similar pattern and is left for the reader. □

The converse to this result is also true, as the following result implies.

Theorem 2.4 *If V is a linear space and B is a basis for it, then $V \cong (K^B)_0$.*

Proof In Proposition 2.3 we established that the function $T : (K^B)_0 \to V$ given by

$$T(f) = \sum_{b \in B} f(b) \cdot b$$

is a bijection. We show now that it is in fact an isomorphism, thus completing the proof. We need to verify that $T(\alpha_1 f + \alpha_2 g) = \alpha_1 T(f) + \alpha_2 T(g)$, which amounts to showing that

$$\sum_{b \in B} (\alpha_1 f + \alpha_2 g)(b) \cdot b = \alpha_1 \sum_{b \in B} f(b) \cdot b + \alpha_2 \sum_{b \in B} g(b) \cdot b,$$

which follows by an immediate computation. □

Remark 2.9 Every linear space V over K is thus isomorphic to a linear space of the form $(K^B)_0$. The latter space is clearly only dependent on the cardinality of the set B, i.e., if X is any set with the same cardinality as B, then $(K^B)_0 \cong (K^X)_0$. In other words, the dimension of a linear space determines it up to an isomorphism. It should be noted however that generally there is no canonical choice for an isomorphism between V and $(K^X)_0$.

In the finite dimensional case we obtain the following corollary.

Corollary 2.2 *Let V be a finite dimensional linear space over the field K. Any choice of a basis $x_1, \ldots, x_n$ for V gives rise to an isomorphism $V \to K^n$. In this way we may identify every vector in V, by means of coordinates, with an n-tuple of scalars.*

In a more general context, and in particular in the infinite dimensional case, where bases are typically not explicitly available, we have the following results.

Corollary 2.3 *If V and W are linear spaces over the same field K and they have the same dimension, then they are isomorphic.*

Proof By Remark 2.9, if we let X be a set of cardinality equal to the common dimensions of V and W, then we obtain an isomorphisms $T_1 : (K^X)_0 \to V$ and an isomorphism $T_2 : (K^X)_0 \to W$. It follows that $T_2 \circ T_1^{-1} : V \to W$ is an isomorphism, showing that $V \cong W$, as claimed. □

Combining Proposition 2.9 and Corollary 2.3, the discussion above is summarized as follows.

Theorem 2.5 *Two linear spaces over the same field are isomorphic if, and only if, they have the same dimension.*

Remark 2.10 This result is an example of what is known as a rigidity phenomenon. Rigidity refers to a situation where two structures are essentially the same given that they have essentially the same substructure of some kind, typically of a much coarser nature than the original structures. In this case, the rigidity of linear spaces over a fixed field K is that the dimension, i.e., the cardinality of a basis, suffices to determine the linear space, up to an isomorphism.

Example 2.20 Recall the linear space P_n of polynomial functions from Example 2.6. We can easily show that $P_n \cong \mathbb{R}^{n+1}$ by constructing an isomorphism between the two spaces. Indeed, referring here to a typical element in $\mathbb{R}^{n+1}$ by $(a_0, \ldots, a_n)$, let

$$T : \mathbb{R}^{n+1} \to P_n$$

be given by

$$T(a_0, \ldots, a_n) = a_n t^n + \cdots + a_1 t + a_0.$$

It is a trivial matter to verify that T is a linear operator, clearly bijective, and thus an isomorphism. Similarly, the reader is invited to show that $c_{00} \cong P$. By the discussion above, the existence of an isomorphism (but not any explicit isomorphism) could have been deduced by noting that both $\mathbb{R}^{n+1}$ and P_n, as linear spaces over $\mathbb{R}$, have dimension $n + 1$. Similarly, both c_{00} and P have countably infinite dimension and are thus isomorphic.

Remark 2.11 Theorem 2.5 tells us that the dimension of a linear space essentially is all that one needs to know about a linear space in order to study it (since after all, isomorphic linear spaces are essentially the same). It is thus tempting to, once and for all, choose a single representative from each isomorphism class of linear spaces. This certainly suffices for the study of all linear spaces, and seems more economical than having all of this redundancy in the form of multiple isomorphic specimens of linear spaces. However, the example above clearly shows the danger in this approach. While $\mathbb{R}^{n+1}$ and P_n are isomorphic, they have different qualities (at least to us humans). For instance, it is very natural to consider the integral as an operator on polynomials, but not so much on elements in $\mathbb{R}^{n+1}$. The richness of having many different linear spaces, even if they are isomorphic, is a blessing that should not be given up for the sake of a more economical treatment.

Exercises

Exercise 2.21 Consider the linear space c of all convergent sequences of complex numbers (or real numbers, with the obvious adjustments). Prove that the assignment

$$\{x_m\}_{m\geq 1} \mapsto \lim_{m\to\infty} x_m$$

is a linear operator from c to $\mathbb{C}$.

Exercise 2.22 Finish the proof of Theorem 2.3 showing that $\mathrm{Hom}(V, W)$ is a linear space.

Exercise 2.23 Let V and W be finite-dimensional linear spaces over a field K, of dimensions n and m respectively, and let $B_1 = \{v_1, \ldots, v_n\}$ and $B_2 = \{w_1, \ldots, w_m\}$ be fixed bases for V and W respectively. For any vector $x \in V$, we write

$$[x]_{B_1} = (a_1, \ldots, a_n) \in K^n$$

where $a_1, \ldots, a_n$ are the unique scalars such that

$$x = \sum_{k=1}^{n} a_k v_k.$$

The tuple $[x]_{B_1}$ is called the vector of coordinates of x in the basis B_1. Similarly, one defines $[y]_{B_2}$ for all $y \in W$.

1. Given a linear operator $T : V \to W$, prove that there exists a unique matrix $[T]_{B_1,B_2} \in M_{n,m}(K)$ such that

   $$[T(x)]_{B_2} = [T]_{B_1,B_2} \cdot [x]_{B_1}$$

   where on the right-hand-side the '$\cdot$' stands for the ordinary product of a matrix by a vector. The matrix $[T]_{B_1,B_2}$ is called the *representative* matrix of the linear operator T in the given bases.
2. Prove that the assignment $T \mapsto [T]_{B_1,B_2}$, mapping a linear operator to the matrix representing it in the two given bases, is a linear isomorphism

   $$\psi_{B_1,B_2} : \mathrm{Hom}(V, W) \to M_{n.m}(K).$$

3. Suppose now that U is a third linear space with its chosen basis B_3 and that $S : W \to U$ is a linear operator. Prove that $[S \circ T]_{B_1,B_3} = [S]_{B_2,B_3} \cdot [T]_{B_1,B_2}$, where the product on the right-hand-side is the ordinary product of matrices.

We remark that this exercise gives one justification for the definition of the algebraic operations on matrices the way they are defined.

Exercise 2.24 Prove that $\mathbb{R}^n$ and $\mathbb{C}^n$, $n \geq 1$, when considered as linear spaces over $\mathbb{R}$, are not isomorphic.

Exercise 2.25 Prove that $\mathbb{R}^\infty \cong \mathbb{C}^\infty$, when considered as linear spaces over $\mathbb{R}$.

Exercise 2.26 A linear operator $T : V \to V$ is *nilpotent* if there exists an $m \geq 1$ such that T^m, the m-fold composition of T with itself, is the zero operator $0 : V \to V$. Prove that if $T : V \to V$ is nilpotent, then $\mathrm{id}_V - T$ is an isomorphism.

Exercise 2.27 Let $T : V \to V$ be an operator with the property that for all $x \in V$ there exists an $m \geq 1$ such that $T^m(x) = 0$. Prove that if V is finite-dimensional, then T is nilpotent, but if V is infinite-dimensional, then T need not be nilpotent.

Exercise 2.28 Prove that for every set $\mathscr{B}$ there exists a linear space with $\mathscr{B}$ as a basis (and consequently for any cardinality κ there exists a linear space whose dimension is κ).

Exercise 2.29 Let $f : \mathbb{R} \to \mathbb{R}$ be a continuous function with $f(x + y) = f(x) + f(y)$ for all $x, y \in \mathbb{R}$. Prove that there exists some $a \in \mathbb{R}$ such that $f(x) = ax$ for all $x \in \mathbb{R}$.

Exercise 2.30 Use a Hamel basis for $\mathbb{R}$ as a linear space over $\mathbb{Q}$ to construct a function $f : \mathbb{R} \to \mathbb{R}$ satisfying $f(x + y) = f(x) + f(y)$ for all $x, y \in \mathbb{R}$, which is not of the form $f(x) = ax$ for any $a \in \mathbb{R}$.

2.4 Subspaces, Products, and Quotients

Subspaces arise naturally as portions of a given linear space that inherit a linear space structure from the ambient linear space. We show that any linear operator gives rise to certain subspaces, its kernel and its image, and we show how to combine two linear spaces to form their product space. We also show how to eliminate a subspace so as to obtain a quotient space. The quotient construction is related to the concept of complementary subspaces, which is also introduced and the connection made explicit.

2.4.1 Subspaces

Given a subset $A \subseteq V$ of a linear space V the operations of addition and scalar multiplication, when restricted to vectors in A always yield elements in the ambient space V. We are interested in the case where the results of these operations always yield elements in A.

Definition 2.7 A subset $A \subseteq V$ is said to be *closed under addition* if

$$x + y \in A$$

for all vectors $x, y \in A$. Similarly, A is said to be *closed under scalar multiplication* if

$$\alpha x \in A$$

for all vectors $x \in A$ and scalars $\alpha \in K$. The set A is called a *linear subspace* (or simply *subspace* if the linear context is evident) of V if $A \neq \emptyset$ and A is closed under addition and scalar multiplication. A subspace A is called a *proper subspace* if it is properly contained in the ambient space V.

Remark 2.12 If A is a linear subspace of V, then A, with vector addition and scalar multiplication induced from V, is a linear space. Indeed, the verification of each of the axioms follows the same pattern. For instance,

$$x + y = y + x$$

holds for all $x, y \in A$ since the same equality holds for all vectors x and y in the ambient space V. Moreover, notice that the zero vector 0 is always an element in A and is the zero vector of A. Indeed, $0 = 0 \cdot x$ holds for all $x \in A$, and since A is required to be non-empty, at least one $x \in A$ exists.

Notice that the subset $\{0\}$ consisting of just the zero vector is always a subspace, called the *trivial subspace* of V. Another immediate example of a subspace of V is V itself, i.e., a non-proper subspace. An immediate property of subspaces is their transitive property, that is if $A \subseteq B \subseteq V$, then if B is a subspace of V and A is a subspace of B, then A is a subspace of V.

Example 2.21 Consider the linear space $\mathbb{R}^2$ and its subset A consisting only of the vectors of the form $(x, 0) \in \mathbb{R}^2$. Clearly, A is non-empty, closed under addition, and closed under scalar multiplication, and hence is a subspace of $\mathbb{R}^2$. More generally, a line $l = \{tx \mid t \in \mathbb{R}\}$, with $x \in \mathbb{R}^2$ a non-zero vector, is a linear subspace of $\mathbb{R}^2$. These linear subspaces, together with the subspaces $\{0\}$ and $\mathbb{R}^2$, exhaust all of the linear subspaces of $\mathbb{R}^2$. Similarly, linear subspaces of $\mathbb{R}^n$, for larger values of n, correspond to the origin, to lines through the origin, to planes through the origin, and to hyper-planes through the origin. Notice the following subtle point. The field $\mathbb{R}$, when viewed as a linear space over itself, is obviously isomorphic to $\mathbb{R}^1$, that is to $\mathbb{R}^n$ where $n = 1$. However, even though the linear space $\mathbb{R} \cong \mathbb{R}^1$ may naturally be identified with either the **X**-axis or the **Y**-axis in $\mathbb{R}^2$, formally speaking, $\mathbb{R}$ itself (or $\mathbb{R}^1$) is not a subspace of $\mathbb{R}^2$. In fact, it is not even a subset of $\mathbb{R}^2$. More generally, $\mathbb{R}^n$ is a subspace of $\mathbb{R}^m$ if, and only if, $n = m$, in which case the two spaces coincide. When $n \leq m$ the space $\mathbb{R}^n$ may be identified in finitely many ways with various subspaces of $\mathbb{R}^m$, but that is a different story. An analogous discussion can be given for linear spaces over $\mathbb{C}$.

Example 2.22 Consider the linear space $\mathbb{C}^\infty$ from Example 2.3 and the linear spaces c of convergent sequences, c_0 of sequences that converge to 0, and c_{00} of sequences that are eventually 0 (see Example 2.4). One clearly has that

$$c_{00} \subset c_0 \subset c \subset \mathbb{C}^\infty$$

and in fact c_{00} is a linear subspace of c_0, which in turn is a linear subspace of c, which is a linear subspace of $\mathbb{C}^\infty$. The verification is immediate.

Other subspaces of $\mathbb{C}^\infty$ include, for each $n \geq 1$, the set of all vectors of the form $(x_1, \ldots, x_n, 0, 0, \ldots)$. This space is clearly isomorphic to $\mathbb{C}^n$, as are infinitely many other subspaces of $\mathbb{C}^\infty$, as the reader is asked to verify. In any event, $\mathbb{C}^n$ is not itself a subspace of $\mathbb{C}^\infty$.

Example 2.23 Consider the linear space P_n (Example 2.6) consisting of polynomial functions with real coefficients of degree not exceeding n, and P, the linear space of all polynomials with real coefficients. For all $n \leq m$, it holds that $P_n \subset P_m \subset P$, and it follows easily that P_n is a subspace of P_m, which in turn is a subspace of P. We thus obtain the tower of proper subspace inclusions

$$P_0 \subset P_1 \subset P_2 \subset \cdots \subset P_n \subset \cdots \subset P.$$

Example 2.24 With reference to Example 2.7, the subset of $C([a, b], \mathbb{R})$ consisting of the continuous functions $x : [a, b] \to \mathbb{R}$ which vanish at a and b, that is such that $x(a) = x(b) = 0$, constitutes a linear subspace. On the contrary, the subset of those continuous functions with $x(a) = x(b) = 1$ does not constitute a linear subspace (for instance since it fails to contain the zero vector).

The following family of linear spaces, which are subspaces of $\mathbb{C}^\infty$, is of significant importance.

Example 2.25 (The ℓ_p Spaces). Consider the linear space $\mathbb{C}^\infty$ from Example 2.3 and recall that a sequence $x = (x_1, \ldots, x_k, \ldots) \in \mathbb{C}^\infty$ is said to be *bounded* if there exists some $M \in \mathbb{R}$ such that

$$|x_k| < M$$

for all $k \geq 1$. Let $\ell_\infty \subseteq \mathbb{C}^\infty$ be the set of all bounded sequences. Clearly, the zero vector, namely the constantly zero sequence, is in ℓ_∞, the sum of two bounded sequences is again bounded, and thus in ℓ_∞, and if x is bounded by M, then $\alpha x = (\alpha x_1, \ldots, \alpha x_k, \ldots)$ is bounded by $|\alpha| M$, and thus is in ℓ_∞. It thus follows that ℓ_∞ is a linear subspace of $\mathbb{C}^\infty$.

Now, let $p \geq 1$ and consider the set ℓ_p of all sequences of complex numbers that are *absolutely p-power summable*, that is, all sequences $x \in \mathbb{C}^\infty$ such that

$$\sum_{k=1}^{\infty} |x_k|^p < \infty.$$

Clearly, the constantly zero sequence belongs to ℓ_p, and ℓ_p is evidently closed under scalar multiplication. To show that ℓ_p is a subspace of $\mathbb{C}^\infty$ it thus remains to be shown that ℓ_p is closed under addition, a task we leave to the reader.

We thus obtain a one-parameter family $\{\ell_p\}_{1 \leq p \leq \infty}$ of linear subspaces of $\mathbb{C}^\infty$. In fact, it is easily seen that if $1 \leq p < q \leq \infty$, then $\ell_p \subset \ell_q$ (consider a suitable variation of the harmonic series), and so ℓ_p is a proper linear subspace of ℓ_q.

Example 2.26 Recall from Example 2.8 that $\mathbb{R}$ may be viewed as a linear space over $\mathbb{R}$, and that $\mathbb{C}$ can be viewed as a linear space over $\mathbb{C}$ or as a linear space over $\mathbb{R}$. The set inclusion $\mathbb{R} \subseteq \mathbb{C}$ is of course always valid, but $\mathbb{R}$ is a linear subspace of $\mathbb{C}$ only when both are considered to be linear spaces over $\mathbb{R}$ (or over $\mathbb{Q}$). A subspace relation between two linear spaces can only hold if both linear spaces are over the same field, and so, for instance, $\mathbb{C}^n$ as a linear space over $\mathbb{R}$ is not a linear subspace of $\mathbb{C}^n$ as a linear space over $\mathbb{C}$.

We now address the relationship between subspaces and dimension which, in the finite case, is as one would expect. But the infinite case, as usual, is more subtle.

Theorem 2.6 *For a linear space V of dimension n and a subspace $U \subseteq V$ with dimension m, the following hold.*

1. $m \leq n$.
2. If V is finite dimensional and $m = n$, then $U = V$.

Proof

1. The proof is given in full generality, thus n and m may be infinite cardinals. Let B_V be a basis for V and let B_U be a basis for U. By assumption, the cardinality of B_V is n while the cardinality of B_U is m. Now, by definition of basis, B_U is a linearly independent set of vectors in U, and hence is also linearly independent in V, while B_V is a spanning set in V. Applying Lemma 2.1, it follows that there exists an injection $B_U \to B_V$, and thus $n \leq m$.
2. Assume now that $m = n$ and that V is finite dimensional, namely that $n = m$ is a natural number. We may thus find a basis $\{x_1, \ldots, x_n\}$ of U. In particular these vectors are linearly independent in U, and thus also in V. By Proposition 2.4, it follows that the set $\{x_1, \ldots, x_n\}$ is a basis of V, so its span is V. But the span is also U, and the claim follows.

□

Remark 2.13 Much as in Remark 2.6, the finite dimensionality assumption cannot be avoided. Indeed, the linear space c_{00} from Example 2.4 is, as we saw, of countably infinite dimension but it does contain proper subspaces of the same dimension. For instance, it is easy to verify that the set c_{000} of all sequences in c_{00} whose first term is 0, is a proper subspace of c_{00} even though it is isomorphic to c_{00}, and thus has the same dimension as the ambient space.

2.4.2 Kernels and Images

With any linear operator $T : V \to W$ one can associate a subspace of the domain, called the kernel of T, and a subspace of the codomain, called the image of T. As we show below, the kernel suffices to detect whether or not T is injective.

Definition 2.8 Let $T : V \to W$ be a linear operator. The set

$$\text{Ker}(T) = \{x \in V \mid T(x) = 0\}$$

is called the *kernel* of T.

Theorem 2.7 *The kernel of a linear operator $T : V \to W$ is a linear subspace of V.*

Proof In Proposition 2.5 we already noted that $T(0) = 0$ and thus $0 \in \text{Ker}(T)$. All we need to do now is show that the kernel is closed under vector addition and scalar product. Indeed, if $x, y \in \text{Ker}(T)$, then

$$T(x + y) = T(x) + T(y) = 0 + 0 = 0$$

and thus $x + y \in \text{Ker}(T)$. Similarly, for all $\alpha \in K$, if $x \in \text{Ker}(T)$, then

$$T(\alpha x) = \alpha T(x) = \alpha \cdot 0 = 0$$

and thus $\alpha x \in \text{Ker}(T)$ and the proof is complete. □

Theorem 2.8 *A linear operator $T : V \to W$ is injective if, and only if, its kernel is trivial, i.e.,* $\text{Ker}(T) = \{0\}$.

Proof If T is injective, then since we already know that $T(0) = 0$, it follows that $T(x) = 0$ implies $x = 0$, and thus $\text{Ker}(T) = \{0\}$, and is thus trivial. Conversely, suppose that the kernel of T is trivial and suppose that

$$T(x) = T(y)$$

for some $x, y \in V$. Then

$$T(x - y) = T(x) - T(y) = 0$$

and thus

$$x - y \in \text{Ker}(T).$$

But then $x - y = 0$, and so $x = y$, showing that T is injective. □

Another subspace naturally obtained from a linear operator is its image.

Definition 2.9 Let $T : V \to W$ be a linear operator. The set $\mathrm{Im}(T) = \{T(x) \mid x \in V\}$, namely the set-theoretic image of T, is called the *image* of T.

Proposition 2.10 *The image of a linear operator $T : V \to W$ is a linear subspace of W.*

Proof Obviously, $T(V)$ is not empty. Further, if $x, y \in \mathrm{Im}(T)$, then $x = T(x')$ and $y = T(y')$ for some vectors $x', y' \in V$. It then follows that

$$x + y = T(x') + T(y') = T(x' + y') \in \mathrm{Im}(T).$$

Similarly, one shows that $\alpha x \in \mathrm{Im}(T)$ for all $x \in \mathrm{Im}(T)$ and all scalars $\alpha \in K$. □

The following result, which the reader may recognize as the rank-nullity theorem, has important consequences for finite dimensional linear spaces.

Lemma 2.3 *The equality*

$$\dim(V) = \dim(\mathrm{Ker(T)}) + \dim(\mathrm{Im(T)})$$

holds for all linear operators $T : V \to W$ between finite dimensional linear spaces.

Proof We present the general strategy, inviting the reader to fill-in the details. Choose a basis $\{x_1, \ldots, x_k\}$ for $\mathrm{Ker}(T)$. Then augment this basis by (if needed) adding vectors to it, so as to obtain a basis $\{x_1, \ldots, x_k, y_1, \ldots, y_m\}$ of V and show that $\{T(y_1), \ldots, T(y_m)\}$ is a basis for $\mathrm{Im}(T)$. □

Remark 2.14 The proof above can be adapted to obtain a similar result for infinite dimensional linear spaces, provided one handles cardinal arithmetic with care, of course. However, such a result is not of great importance and we thus avoid its details.

Corollary 2.4 *For a linear operator $T : V \to W$ between finite dimensional linear spaces of equal dimension, the conditions*

- *T is injective*
- *T is surjective*
- *T is bijective*

are equivalent.

Proof We have seen that T is injective if, and only if, its kernel is trivial, namely precisely when $\dim(\mathrm{Ker}(T)) = 0$. It follows then that T is injective if, and only if, $\dim(V) = \dim(\mathrm{Im}(T))$. But since $\dim(V) = \dim(W)$, it follows by Theorem 2.6, that $\dim(V) = \dim(\mathrm{Im}(T))$ holds if, and only if, $\mathrm{Im}(T) = W$, in other words, precisely when T is surjective. This completes the verification of the non-trivial arguments in the proof. □

2.4.3 Products and Quotients

We now present the product construction for two linear spaces V and W over the same field K. We endow the set $V \times W$ with an addition operation and a scalar product that turn it into a linear space, as follows. Given $(x_1, x_2), (y_1, y_2) \in V \times W$, and $\alpha \in K$, the addition operation is given by

$$(x_1, x_2) + (y_1, y_2) = (x_1 + y_1, x_2 + y_2)$$

and the scalar product operation is given by

$$\alpha(x_1, x_2) = (\alpha x_1, \alpha x_2).$$

Theorem 2.9 *For linear spaces V and W over the same field K, the set $V \times W$, when endowed with addition and scalar multiplication as above, is a linear space over the field K.*

Proof The verification of the linear space axioms is straightforward. For instance, the additively neutral element is $(0, 0)$ since

$$(x_1, x_2) + (0, 0) = (x_1 + 0, x_2 + 0) = (x_1, x_2)$$

for all $(x_1, x_2) \in V \times W$. We leave the rest of the verification to the reader. □

Obviously, one can similarly define the product $V_1 \times \cdots \times V_n$ of any finite number of linear spaces, and even the product of any collection of linear spaces. It is also evident, upon inspection of the linear structure of the linear space $\mathbb{R}^n$, that $\mathbb{R}^n$ is the n-fold product of $\mathbb{R}$ with itself, where $\mathbb{R}$ is viewed as a linear space over itself. Similarly, $\mathbb{C}^n$ is the n-fold product of $\mathbb{C}$ with itself.

We now turn to consider the quotient construction of a linear space by a linear subspace. This construction is, in some sense, the reversal of taking the product of linear spaces. We refer the reader to Sect. 1.2.15 of the Preliminaries for basic facts about equivalence relations.

Definition 2.10 Let V be a linear space and U a subspace of it. Two elements $x_1, x_2 \in V$ are said to be *equivalent modulo U* if

$$x_1 - x_2 \in U$$

that is, if

$$x_1 = x_2 + u$$

for some $u \in U$. This relation is denoted by $x_1 = x_2 (\text{mod } U)$, by $x \equiv_U y$, or simply by $x \equiv y$ if U is evident.

We now show that $\equiv$ is an equivalence relation on V. Indeed, for all $x, y, z \in V$

$$x - x = 0 \in U,$$

and thus

$$x \equiv x,$$

so that $\equiv$ is reflexive. Further,

$$x \equiv y \implies x - y \in U \implies y - x = (-1) \cdot (x - y) \in U \implies y \equiv x,$$

and thus $\equiv$ is symmetric. Finally,

$$x - y, y - z \in U \implies x - z = (x - y) + (y - z) \in U \implies x \equiv z,$$

and so $\equiv$ is transitive.

Recall that the equivalence class of any vector $x \in V$ is the set

$$[x] = \{y \in V \mid x \equiv y\}.$$

In other words, it is the set of all vectors of the form $x + u$, where $u \in U$. For that reason the equivalence class $[x]$ is also denoted by

$$x + U.$$

It follows from the general theory of equivalence relations that $\{x + U \mid x \in V\}$ is a partition of V.

Example 2.27 As a simple example, consider the usual 2-dimensional plane $\mathbb{R}^2$ and its subspace $\mathbf{Y}$ consisting of the ordinates. Let the element $\mathbf{r} \in \mathbb{R}^2$ be the vector $O - P$ from the origin to the point P in the plane. The equivalence class $[\mathbf{r}]_{\mathbf{Y}}$, formed by the vectors of $\mathbb{R}^2$ equivalent mod $\mathbf{Y}$ to $\mathbf{r}$, is clearly given by all vectors $O - P_i$ with P_i lying on the parallel line to the $\mathbf{Y}$-axis passing through the point P; all these vectors are equivalent *mod* $\mathbf{Y}$ to each other.

The fact that the entire 2-dimensional space is partitioned by all such equivalence classes is simply the fact that $\mathbb{R}^2$ is the disjoint union of all the lines parallel to the $\mathbf{Y}$ axis, which are precisely the translates $x + \mathbf{Y}$ of $\mathbf{Y}$.

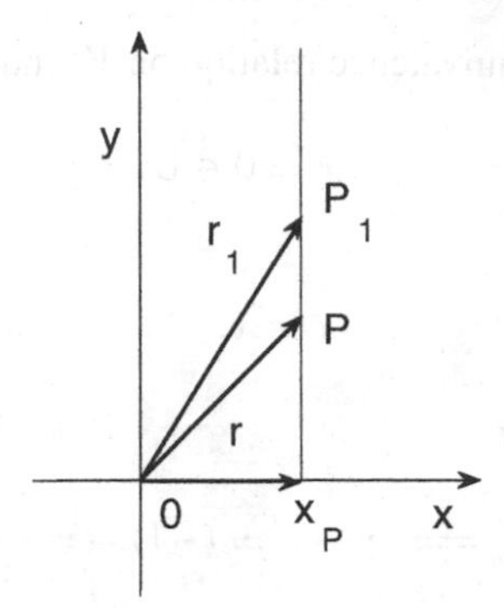

We now turn to investigate the quotient set $V/U = \{x+U \mid x \in V\}$ of equivalence classes modulo U. It is natural to introduce an addition and a scalar product on V/U by the formulas

$$(y+U)+(z+U)=(y+z)+U \qquad \text{and} \qquad \alpha(y+U)=(\alpha y)+U,$$

but we must first verify that these operations are well-defined, namely that they do not depend on the chosen representatives. Indeed, for all $x, y, x', y' \in V$, we need to show that if

$$x+U=x'+U \;\text{ and }\; y+U=y'+U,$$

then

$$(x+y)+U=(x'+y')+U.$$

But the former implies that

$$x'-x \in U \;\text{ and }\; y'-y \in U$$

while the latter will be implied by showing that $(x'+y')-(x+y) \in U$. And indeed, since U is closed under addition,

$$(x'+y')-(x+y)=(x'-x)+(y'-y) \in U.$$

A similar argument shows that the scalar product above is well-defined and thus we see that the quotient set V/U of equivalence classes is naturally endowed with an addition operation and a scalar product operation.

Theorem 2.10 *Let V be a linear space and $U \subseteq V$ a subspace. The quotient set V/U of equivalence classes modulo U, with the operations of addition and scalar multiplication as given above, is a linear space whose zero vector is U. Moreover, the canonical projection $\pi : V \to V/U$ given by $\pi(x) = x+U$ is a surjective linear operator.*

Proof To see that $U = 0 + U$ behaves neutrally with respect to addition, note that for all $x + U \in V/U$

$$(x + U) + U = (x + U) + (0 + U) = (x + 0) + U = x + U.$$

The rest of the verification of the linear space axioms, as well as the claims about the canonical projection, follow similarly and are left for the reader. □

Definition 2.11 The linear space V/U constructed in Theorem 2.10 is called the *quotient space of* V *with respect to* U, or simply as V *modulo* U.

Continuing the example above, $\mathbb{R}^2/\mathbf{Y}$, which geometrically can be thought of as the space of all vertical lines in $\mathbb{R}^2$, is a linear space. Another, more complicated, example is c_0/c_{00}, the quotient of the space c_0 of all converging sequences of (say) real numbers, by the subspace of all eventually 0 sequences. The quotient may be thought of as the space of all sequences modulo finite changes. In analysis it is well known that the limit of a sequence is insensitive to finite changes in the sequence, and so it is precisely this quotient that is of interest already in elementary analysis.

2.4.4 Complementary Subspaces

Contemplating the quotient space associated to Example 2.27, one realizes that $\mathbb{R}^2/\mathbf{Y}$ is isomorphic to $\mathbf{X}$, and that under this isomorphism, the canonical projection $\pi : \mathbb{R}^2 \to \mathbb{R}^2/\mathbf{Y}$ is nothing but the projection of $\mathbb{R}^2$ onto the $\mathbf{X}$ axis. This observation generalizes to any quotient space construction, as we now show.

Definition 2.12 Two subspaces U and W of a given linear space V are said to be *complementary*

- if the only vector they have in common is the zero vector, in other words if $U \cap W = \{0\}$; and
- if $U + W = V$; that is, given $x \in V$, there exist $u \in U$ and $w \in W$ with $x = u + w$.

Remark 2.15 The decomposition above of an element $x \in V$ as the sum $x = u + w$ is in fact unique. Indeed, if $x = u + w$ and also $x = u' + w'$, with $u, u' \in U$ and $w, w' \in W$, then

$$0 = x - x = (u + w) - (u' + w') \implies u - u' = w' - w.$$

But $u - u' \in U$ and $w' - w \in W$, and since the only element in $U \cap W$ is the zero vector, it follows that $u - u' = 0 = w' - w$, and therefore, $u = u'$ and $w = w'$. This unique decomposition is said to express V as a *direct sum* of its subspaces U and W, a fact denoted by $V = U \oplus W$.

The following example shows that a given subspace $U \subseteq V$ may admit more than one complementary subspace.

Example 2.28 (Continuing Example 2.27) In $\mathbb{R}^2$ any straight line **N** passing through the origin and not coincident with **Y** is a complementary subspace, as can easily be verified by elementary means.

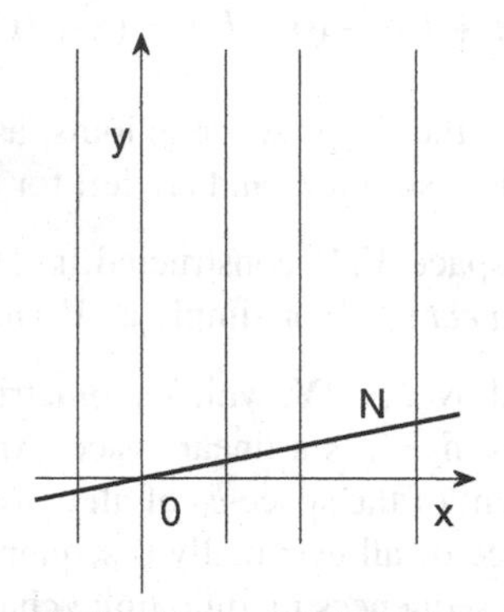

We may now present the main theorem relating complementary subspaces and the quotient construction.

Theorem 2.11 *Given a linear space V and two complementary subspaces U and W, the quotient space V/U is naturally isomorphic to W (the meaning of 'naturally' is discussed below).*

Proof Recall from Theorem 2.10 that the canonical projection $\pi : V \to V/U$, given by $\pi(x) = x + U$, is a linear operator. The result will be established by showing that the restriction of π to $W \subseteq V$ is an isomorphism between W and V/U. All we have to do is show that the restriction is a bijection. For injectivity we use Theorem 2.8 and show the kernel of the restriction is trivial. For that, suppose that $\pi(w) = 0$ for some $w \in W$, that is (recall that the zero vector in V/U is U)

$$w + U = U$$

and so $w \in U$. But since

$$U \cap W = \{0\},$$

it follows that $w = 0$, as needed. For surjectivity, let $y + U$ be an arbitrary element in V/U. Writing $y = u + w$, with $u \in U$ and $w \in W$, we get that

$$y + U = (u + w) + U = (u + U) + (w + U) = w + U = \pi(w). \qquad \square$$

Remark 2.16 The meaning of the isomorphism above being *natural* (or *canonical*) is that its construction does not depend on any choice of basis. This is an important observation since, as we saw, in infinite dimensional spaces it may be impossible to actually describe any basis.

Corollary 2.5 *Given a linear space V and a subspace $U \subseteq V$, any two subspaces W_1 and W_2 that are complementary to U are naturally isomorphic.*

Proof Each of W_1 and W_2 is naturally isomorphic to V/U. □

In the case of complementary spaces, the stated isomorphism is independent of a basis but it is dependent on the ability to express the ambient space as the direct sum of the given subspace and each of its complements. If the latter can be done explicitly, then an explicit formula for the isomorphism will emerge.

Exercises

Exercise 2.31 Let V be a linear space and $S \subseteq V$ a subset. Prove that S is a linear subspace of V if, and only if, S is closed under linear combinations, i.e., if

$$\sum_{k=1}^{m} \alpha_k s_k \in S$$

for all $\alpha_1, \ldots, \alpha_m \in K$ and $s_1, \ldots, s_m \in S$, and $m \geq 0$. Here we adopt the convention that the empty sum, i.e., the case $m = 0$, is equal to 0.

Exercise 2.32 Prove Lemma 2.3: If $T : V \to W$ is a linear operator between finite dimensional linear spaces, then

$$\dim(V) = \dim(\mathrm{Ker(T)}) + \dim(\mathrm{Im(T)}).$$

Exercise 2.33 Let V be a linear space and let $\{S_i\}_{i \in I}$ be a family of subspaces of V. Prove that the intersection

$$S = \bigcap_{i \in I} S_i$$

is a linear subspace of V. On the other hand show that if $S_1, S_2 \subseteq V$ are two linear subspaces of V, then $S_1 \cup S_2$ is a linear subspace of V if, and only if, either $S_1 \subseteq S_2$ or $S_2 \subseteq S_1$.

Exercise 2.34 (Refer to Example 2.25) Prove that ℓ_p, for $1 \leq p \leq \infty$, is a linear space and that if $1 \leq p < q \leq \infty$, then $\ell_p \subset \ell_q$.

Exercise 2.35 Let V and W be two linear spaces over the same field K. Prove that $V \times W \cong W \times V$. Can equality ever hold?

Exercise 2.36 Let V be a linear space and $U \subseteq V$ a linear subspace. Prove that the operation

$$\alpha(x + U) = \alpha x + U,$$

for all vectors $x \in V$ and scalars $\alpha \in K$, is well-defined. That is, show that it is independent of the choice of representative for the equivalence class $x + U$.

Exercise 2.37 Let V be a linear space. Prove that $V/V \cong 0$, where 0 denotes any linear space with a single element, necessarily its zero vector. Show also that $V/\{0\} \cong V$, where $\{0\}$ is the trivial subspace of V. Is it possible that any of these isomorphisms is an actual equality?

Exercise 2.38 Let V be a linear space and $U \subseteq V$ a linear subspace. Prove that the canonical projection $\pi : V \to V/U$ given by

$$\pi(x) = x + U$$

is a surjective linear operator.

Exercise 2.39 Let V be a linear space over K and $U \subseteq V$ a subspace. Consider the diagram

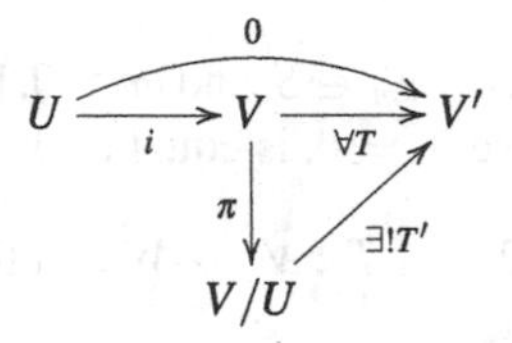

where $i : U \to V$ is the inclusion function, $\pi : V \to V/U$ is the canonical projection to the quotient space, and V' is an arbitrary linear space over K. The rest of the diagram deciphers as follows: Prove that for all linear operators $T : V \to V'$ with the property that $T(x) = 0$ for all $x \in U$, there exists a unique linear operator $T' : V/U \to V'$ such that $T' \circ \pi = T$. (This exercise is establishing what is known as the universal property of the quotient construction)

Exercise 2.40 Let V and W be two linear spaces over the same field K, and consider the product $V \times W$. Show that

1. $U = \{(x, 0) \mid x \in V\}$ is a linear subspace of $V \times W$ which is isomorphic to V. We denote U more suggestively by $V \times \{0\}$.
2. Prove that $(V \times W)/(V \times \{0\}) \cong W$.

2.5 Inner Product Spaces and Normed Spaces

So far in our treatment of linear spaces two geometric aspects are missing, namely angles between vectors and the length of a vector. This situation is unavoidable since in general linear spaces it need not be possible to coherently define any of these notions. In the presence of extra structure, these notions become available, as we discuss below.

2.5.1 Inner Product Spaces

The reader is most likely familiar with the *standard inner product* on $\mathbb{R}^n$, i.e.,

$$\langle x, y \rangle = \sum_{k=1}^{n} x_k y_k.$$

The definition of a general inner product space is then a generalization of certain properties the standard inner product has. We will not pause here to motivate this definition any further, except for two comments. The first comment is that the results below will show that an inner product allows one (at least in the real case) to speak of angles between vectors, thus obtaining retrospective justification for the axioms. The second comment pertains to the non-arbitrary nature of the standard inner product; the reader is invited to discover it from elementary geometry, i.e., the law of cosines.

Definition 2.13 *(Inner Product Space)* Let V be a linear space over K, where K is either $\mathbb{R}$ or $\mathbb{C}$. Assume that a function $\langle -, - \rangle : V \times V \to K$ is given. The function $\langle -, - \rangle$ is said to be an *inner product* if for all $x, y, z \in V$ and $\alpha, \beta \in K$, the following conditions hold.

1. *Conjugate symmetry* or *Hermitian symmetry*, i.e., $\langle y, x \rangle = \overline{\langle x, y \rangle}$, where $\overline{w}$ denotes the complex conjugate of w.
2. Linearity in the second argument, i.e., $\langle x, \alpha y + \beta z \rangle = \alpha \langle x, y \rangle + \beta \langle x, z \rangle$.
3. The inner product is *positive definite*, i.e., $\langle x, x \rangle \geq 0$ for all $x \in V$, and $\langle x, x \rangle = 0$ implies that $x = 0$, the zero vector. In particular, we then introduce the *norm* of x to be $\|x\| = \sqrt{\langle x, x \rangle}$.

The linear space V together with the function $\langle -, - \rangle$ is then called an *inner product space*. To emphasize that $K = \mathbb{R}$ we use the term *real inner product space*, while the case $K = \mathbb{C}$ is given the term *complex inner product space*. The scalar $\langle x, y \rangle$ is the *inner product* of the given vectors (in that order!).

Remark 2.17 Notice that when $K = \mathbb{R}$, conjugate symmetry reduces to *symmetry*, that is

$$\langle y, x \rangle = \langle x, y \rangle,$$

and the inner product is also linear in its first argument, namely,

$$\langle \alpha x + \beta y, z \rangle = \alpha \langle x, z \rangle + \beta \langle y, z \rangle.$$

In the complex case one generally has that

$$\langle \alpha x + \beta y, z \rangle = \overline{\alpha} \langle x, z \rangle + \overline{\beta} \langle y, z \rangle,$$

as follows easily from conjugate symmetry and linearity in the second argument. In both real and complex inner product spaces, the inner product is additive in each argument. We mention as well that in mathematical circles it is common to write (x, y) instead of $\langle y, x\rangle$, that is to define an inner product as being linear in the first argument rather than the second one. Obviously, the difference is only cosmetic.

Example 2.29 Consider the linear space $\mathbb{R}^n$ and the definition

$$x \cdot y = (x_1, \dots, x_n) \cdot (y_1, \dots, y_n) = \sum_{k=1}^{n} x_k y_k.$$

With this definition it is easily seen that $\mathbb{R}^n$ becomes a real inner product space. Similarly, considering $\mathbb{C}^n$, defining

$$x \cdot y = (x_1, \dots, x_n) \cdot (y_1, \dots, y_n) = \sum_{k=1}^{n} \overline{x_k} y_k$$

endows $\mathbb{C}^n$ with the structure of a complex inner product space. Notice that for $x, y \in \mathbb{R}^\infty$ the 'obvious definition'

$$x \cdot y = (x_1, \dots) \cdot (y_1, \dots) = \sum_{k=1}^{\infty} x_k y_k$$

fails to endow $\mathbb{R}^\infty$ with the structure of an inner product space, simply because the sum may fail to converge, and thus this is not even a function.

Example 2.30 Consider the space $C(I, \mathbb{R})$ of continuous real valued functions on the interval $I = [a, b]$. The familiar properties of the integral show at once that defining, for all $x, y \in C(I, \mathbb{R})$,

$$\langle x, y\rangle = \int_a^b dt\, x(t)y(t)$$

endows $C(I, \mathbb{R})$ with the structure of a real inner product space. With the proper conjugation in the integral above, the space $C(I, \mathbb{C})$ of continuous complex valued functions becomes a complex inner product space.

Before we present any further examples let us explore some elementary, and not so elementary, general results.

Proposition 2.11 *For all vectors x, y in an inner product space V (either real or complex)*

$$\|x + y\|^2 \le \|x\|^2 + 2|\langle x, y\rangle| + \|y\|^2.$$

Proof Expanding $\|x+y\|^2 = \langle x+y, x+y\rangle$ we see that

$$\langle x+y, x+y\rangle = \langle x, x+y\rangle + \langle y, x+y\rangle = \langle x, x\rangle + \langle x, y\rangle + \langle y, x\rangle + \langle y, y\rangle$$

and thus the claim will follow by showing that

$$\langle x, y\rangle + \langle y, x\rangle \le 2 \cdot |\langle x, y\rangle|,$$

which is immediate if V is a real inner product space. In the complex case suppose that $\langle x, y\rangle = u + iv$. Then

$$\langle x, y\rangle + \langle y, x\rangle = \langle x, y\rangle + \overline{\langle x, y\rangle} = 2 \cdot u$$

and since

$$|u| = \sqrt{u^2} \le \sqrt{u^2 + v^2} = |u + iv| = |\langle x, y\rangle|,$$

the result follows. □

The geometric interpretation of the norm $\|x\|$ is that it is the length of the vector x. As for the geometric meaning of the inner product, notice that any non-zero vector $x \in V$ can be *normalized* by defining $\hat{x} = x/\|x\|$, a vector of *unit* length. For any vector x of unit length and an arbitrary vector y, the scalar $\langle x, y\rangle$ is interpreted as the *component* of the vector y in the direction x. Consequently, we make the following definition.

Definition 2.14 Let V be an inner product space. Two vectors $x, y \in V$ are said to be *orthogonal* or *perpendicular* if $\langle x, y\rangle = 0$.

Theorem 2.12 (Pythagoras' Theorem) *The equality*

$$\|x+y\|^2 = \|x\|^2 + \|y\|^2$$

holds for all orthogonal vectors x and y in an inner product space.

Proof $\|x+y\|^2 = \langle x+y, x+y\rangle = \|x\|^2 + \langle x, y\rangle + \langle y, x\rangle + \|y\|^2 = \|x\|^2 + \|y\|^2$.

□

2.5.2 The Cauchy-Schwarz Inequality

The following inequality is among the most important inequalities in mathematics. It has numerous uses, some of which we will see immediately and some later on.

Theorem 2.13 (Cauchy-Schwarz Inequality) *The inequality*

$$|\langle x, y\rangle| \leq \|x\|\|y\|$$

holds for all vectors x, y in an inner product space V.

Proof For simplicity let us assume V is a real inner product space (only a small adaptation is needed for the complex case). The inequality is trivial when $x = 0$ and so we proceed assuming that $x \neq 0$. Let $\hat{x} = x/\|x\|$ be the normalization of x, so in particular $\|\hat{x}\| = 1$. Dividing the desired inequality by $\|x\|$ we see that we need to establish the inequality

$$|\langle \hat{x}, y\rangle| \leq \|y\|$$

(and so the Cauchy-Schwarz Inequality acquires the interpretation that the absolute value of a component of y in a given unit direction sets a lower bound on the length of y). Rewriting y to identify its component in the $\hat{x}$ direction we obtain

$$y = y_{\hat{x}} + (y - y_{\hat{x}})$$

where $y_{\hat{x}} = \langle \hat{x}, y\rangle\hat{x}$. Since

$$\langle \hat{x}, y - y_{\hat{x}}\rangle = \langle \hat{x}, y\rangle - \langle \hat{x}, y_{\hat{x}}\rangle = \langle \hat{x}, y\rangle - \langle \hat{x}, y\rangle\langle \hat{x}, \hat{x}\rangle = 0$$

it follows that $y_{\hat{x}}$ and $y - y_{\hat{x}}$ are orthogonal. Pythagoras' Theorem (Theorem 2.12) now yields

$$\|y\|^2 = \|y_{\hat{x}}\|^2 + \|y - y_{\hat{x}}\|^2 \geq \|y_{\hat{x}}\|^2 = \langle \hat{x}, y\rangle^2.$$

Extracting square roots completes the proof. □

Corollary 2.6 *Since we now know that*

$$-1 \leq \frac{\langle x, y\rangle}{\|x\|\|y\|} \leq 1$$

for all non-zero vectors x and y in an arbitrary real inner product space, it follows that

$$\theta = \arccos \frac{\langle x, y\rangle}{\|x\|\|y\|}$$

is defined. Noting that

$$\langle x, y\rangle = \cos\theta\,\|x\|\|y\|$$

we define θ to be the angle between the vectors x and y.

We had just seen that a consequence of the Cauchy-Schwarz inequality is that, in a real inner product space, angles between vectors are a well-defined notion. There are many more consequences of the Cauchy-Schwarz inequality, among which we mention one. The Heisenberg uncertainty principle in Quantum Mechanics is the result of applying the Cauchy-Schwarz inequality in a certain Hilbert space, suitably constructed. Unfortunately, this result falls slightly short of the scope of this book.

2.5.3 Normed Spaces

In an inner product space every vector can be assigned a norm. A normed space is then an abstraction of certain key properties of this norm. This is useful since, as will be shown below, many linear spaces fail to admit an inner product but do admit the structure of a normed space. In such spaces one may still speak of the length of a vector, but not (necessarily) of angles between vectors.

Definition 2.15 *(Normed Space)* A linear space V with a function $x \mapsto \|x\|$, which associates with every vector $x \in V$ a real number $\|x\|$, called the *norm* of x, is said to be a *normed space* if for all vectors $x, y \in V$ and $\alpha \in K$ the following hold.

1. Positivity, i.e., $\|x\| > 0$ provided $x \neq 0$.
2. Homogeneity, i.e., $\|\alpha x\| = |\alpha|\|x\|$.
3. Triangle inequality, i.e., $\|x + y\| \leq \|x\| + \|y\|$.

The following result is an immediate consequence of the axioms.

Proposition 2.12 *Let V be a normed linear space. For all vectors $x, y \in V$:*

1. $\|x\| = 0$ *if, and only if,* $x = 0$.
2. $\|-x\| = \|x\|$.
3. $\|x - y\| \leq \|x\| + \|y\|$.
4. $|\|x\| - \|y\|| \leq \|x - y\|$.

Proof

1. Using homogeneity, $\|0\| = \|0 \cdot 0\| = 0 \cdot \|0\| = 0$, and together with positivity the claim follows.
2. Using homogeneity, $\|-x\| = \|(-1) \cdot x\| = 1 \cdot \|x\| = \|x\|$.
3. Using (2) and the triangle inequality, $\|x - y\| = \|x + (-y)\| \leq \|x\| + \|y\|$.
4. We need to show that $-\|x - y\| \leq \|x\| - \|y\| \leq \|x - y\|$. Using (3) and the triangle inequality, $\|x\| = \|(x + y) - y\| \leq \|x + y\| + \|y\|$ and similarly $\|y\| \leq \|x\| + \|x - y\|$, as needed. □

The definition of normed space was motivated by certain intuitive properties that the lengths of vectors, as modeled by the norm in an inner product space, satisfy. The next result shows that the inner product formalism indeed gives rise to a normed space, as expected.

Lemma 2.4 *Any inner product space, with its associated notion of norm, is a normed space.*

Proof Recall that the norm in an inner product space is given by $\|x\| = \sqrt{\langle x, x\rangle}$, and condition 1 in the definition of normed space is immediate. For condition 2, we have that

$$\|\alpha x\|^2 = \langle \alpha x, \alpha x\rangle = \alpha\overline{\alpha}\langle x, x\rangle = |\alpha|^2\|x\|^2$$

and the desired equality follows. As for the triangle inequality, by Proposition 2.11

$$\|x + y\|^2 = \langle x + y, x + y\rangle \leq \|x\|^2 + 2|\langle x, y\rangle| + \|y\|^2$$

and by applying the Cauchy-Schwarz inequality we obtain that

$$\|x + y\|^2 \leq \|x\|^2 + 2\|x\|\|y\| + \|y\|^2 = (\|x\| + \|y\|)^2,$$

and the result follows. □

We close the section, and the chapter, by introducing two very important families of normed spaces. The ℓ_p spaces, introduced already in Example 2.25, are spaces of sequences and each carries its own norm, the ℓ_p norm. The second family of spaces, the pre-L_p spaces, allude to the family of L_p spaces and the L_p norm. The pre-L_p spaces are, in a sense, the continuous version of the ℓ_p spaces and the L_p spaces are the completion of the pre-L_p spaces (completions are discussed in Chap. 4).

2.5.4 The Family of ℓ_p Spaces

For the reader's convenience, we repeat here the definition of the ℓ_p spaces, and augment it with the definition of the ℓ_p norm.

Definition 2.16 Let $1 \leq p < \infty$ be a real number. The set ℓ_p consists of all sequences $x = (x_1, \ldots, x_k, \ldots)$ of complex numbers for which

$$\sum_{k=1}^{\infty} |x_k|^p < \infty.$$

The ℓ_p-*norm* of x is

$$\|x\|_p = \left(\sum_{k=1}^{\infty} |x_k|^p\right)^{1/p}.$$

The ℓ_∞ space is the space of all bounded sequences x of complex numbers and the ℓ_∞ norm is

$$\|x\|_\infty = \sup_k |x_k|.$$

Note that it is immediate that $\|x\|_p = 0$ if, and only if, $x = 0$. These spaces are over the field $\mathbb{C}$. By considering sequences of real numbers one obtains similar spaces over the field $\mathbb{R}$ which, ambiguously, are also referred to as ℓ_p spaces.

To further investigate the structure of ℓ_p, we need the following elementary result.

Proposition 2.13 (Young's Inequality) *The inequality*

$$ab \leq \frac{a^p}{p} + \frac{b^q}{q}$$

holds for all positive real numbers p and q satisfying $1/p + 1/q = 1$, and non-negative real numbers a and b, with equality if, and only if, $a^p = b^q$.

Proof Recall the weighted arithmetic-geometric means inequality

$$\sqrt[w]{x_1^{w_1} x_2^{w_2}} \leq \frac{w_1 x_1 + w_2 x_2}{w}$$

which holds for all non-negative real numbers x_1, x_2, and all positive weights w_1, w_2, with $w = w_1 + w_2$. Applying this inequality to the numbers $x_1 = a^p$ and $x_2 = b^q$, with weights $w_1 = 1/p$ and $w_2 = 1/q$, yields the desired inequality. □

The next inequality is used often when manipulating elements of the spaces ℓ_p. It is used below to establish the triangle inequality when proving that each ℓ_p space, with its ℓ_p norm, is a normed space. It is convenient to first introduce the following notation. For any two sequences $x = (x_1, \ldots, x_k, \ldots)$ and $y = (y_1, \ldots, y_k, \ldots)$, let

$$xy = (x_1 y_1, \ldots, x_k y_k, \ldots),$$

namely, the vector of component-wise products of the given vectors.

Theorem 2.14 (Hölder's Inequality) *Given positive real numbers p and q for which $1/p + 1/q = 1$, the inequality*

$$\|xy\|_1 \leq \|x\|_p \cdot \|y\|_q$$

holds for all $x \in \ell_p$ and $y \in \ell_q$.

Proof The case where either $\|x\|_p = 0$ or $\|y\|_q = 0$ is trivial and so we may assume this is not so. By component-wise division of x and y, respectively, by $\|x\|_p$ and $\|y\|_q$, we may assume that $\|x\|_p = \|y\|_q = 1$, namely that

$$\sum_{k=1}^{\infty} |x_k|^p = 1 = \sum_{k=1}^{\infty} |y_k|^q,$$

and we need to show that $\|xy\|_1 \leq 1$, or in other words that

$$\sum_{k=1}^{\infty} |x_k y_k| \leq 1.$$

By Young's Inequality (Proposition 2.13) we have that

$$|x_k y_k| = |x_k||y_k| \leq \frac{|x_k|^p}{p} + \frac{|y_k|^q}{q}$$

and therefore

$$\sum_{k=1}^{\infty} |x_k y_k| \leq \frac{1}{p} + \frac{1}{q} = 1,$$

as required. □

Theorem 2.15 (Minkowski's Inequality) *Let* $1 \leq p < \infty$. *The inequality*

$$\|x + y\|_p \leq \|x\|_p + \|y\|_p$$

holds for all $x, y \in \ell_p$.

Proof Notice that

$$\|x + y\|_p^p = \sum_{k=1}^{\infty} |x_k + y_k|^p = \sum_{k=1}^{\infty} |x_k||x_k + y_k|^{p-1} + \sum_{k=1}^{\infty} |y_k||x_k + y_k|^{p-1}.$$

Applying Hölder's inequality (Theorem 2.14) with the given p and the associated $q = p/(p-1)$, we obtain

$$\sum_{k=1}^{\infty} |x_k||x_k + y_k|^{p-1} \leq (\sum_{k=1}^{\infty} |x_k|^p)^{\frac{1}{p}} (\sum_{k=1}^{\infty} |x_k + y_k|^p)^{1-\frac{1}{p}} = \|x\|_p \frac{\|x + y\|_p^p}{\|x + y\|_p}$$

and a similar inequality for the second summand. It follows that

$$\|x + y\|_p^p \leq (\|x\|_p + \|y\|_p) \frac{\|x + y\|_p^p}{\|x + y\|_p}$$

and, by simplifying, the claimed inequality is established. □

Theorem 2.16 *For all* $1 \leq p \leq \infty$ *the space* ℓ_p *is a normed linear space.*

Proof The case $p = \infty$ is left for the reader, so assume that $p < \infty$. The non-negativity of $\|x\|_p$ is clear from the definition of the norm and it was already noted that it is immediate that $\|x\| = 0$ implies $x = 0$. Homogeneity, is also immediate, since

$$\|\alpha x\|_p^p = \sum_{k=1}^{\infty} |\alpha x_k|^p = |\alpha|^p \|x\|_p^p$$

and finally the triangle inequality

$$\|x + y\|_p \leq \|x\|_p + \|y\|_p$$

is precisely Minkowski's Inequality (Theorem 2.15). □

2.5.5 The Family of Pre-L_p Spaces

With the large family $\{\ell_p\}_{1 \leq p \leq \infty}$ of normed spaces at hand one can put the general theory to use for many sequences. Noticing that $\ell_p \subset \ell_q$ for all $1 \leq p < q \leq \infty$, and with sequences in ℓ_p converging faster to 0 than sequences in ℓ_q, given a bounded sequence one would typically try to identify a suitable p such that the sequence is found in ℓ_p, and proceed to apply the general theory to the problem at hand.

As long as sequences are concerned, the ℓ_p spaces are adequate but, quite often, when modeling a physical problem mathematically one obtains a function rather than a sequence. It is thus desirable to consider suitable normed spaces of functions. As explained above, obtaining the correct analogue of the ℓ_p spaces, namely the L_p spaces, requires some more sophisticated machinery than what we had presented so far, and so at this point we present what we call the pre-L_p spaces.

For the sake of simplicity, let us consider the ambient linear space $C([0, \infty), \mathbb{R})$ of all continuous function $x : [0, \infty) \to \mathbb{R}$ (we note that with the necessary modifications in the coming definitions and proofs, one may alter the domain, as well as consider complex-valued functions). The *pre-L_p* space, for $1 \leq p < \infty$, is the subset K_p of $C([0, \infty), \mathbb{R})$ whose elements are those functions x with

$$\int_0^\infty dt \; |x(t)|^p < \infty$$

where the integral is the usual Riemann integral (since we only consider continuous functions, the theory of Riemann integration is sufficient). The *L_p norm* of $x \in K_p$ is given by

$$\|x\|_p = \left(\int_0^\infty dt \; |x(t)|^p \right)^{1/p}$$

and we immediately note that $\|x\| = 0$ implies $x = 0$ (due to the continuity of x). Lastly, K_∞ is the space of all bounded continuous functions $x : [0, \infty) \to \mathbb{R}$ and the L_∞ norm is given by

$$\|x\|_\infty = \sup_{0 \le t < \infty} \{|x(t)|\}.$$

The rest of the section is devoted to showing that K_p with the L_p norm is a normed space, for each $1 \le p \le \infty$. In fact, the proof is formally identical to the case of the ℓ_p spaces. Indeed, the only ingredient one needs is the following version of Hölder's Inequality.

Theorem 2.17 (Hölder's Inequality) *Given positive real numbers p and q such that* $1/p + 1/q = 1$*, the inequality*

$$\|xy\|_1 \le \|x\|_p \cdot \|y\|_q$$

holds for all $x \in K_p$ *and* $y \in K_q$*, where* xy *is the function* $(xy)(t) = x(t)y(t)$.

Proof An inspection of the proof of Hölder's Inequality for sequences reveals that it was obtained by a point-wise application of Young's Inequality as well as properties of summation which are well-known to hold for integration as well, and so the same argument can be used to establish Hölder's Inequality for functions. □

The details of the proof of the following result are now formally identical to the proof of Theorem 2.16

Theorem 2.18 *For all* $1 \le p \le \infty$ *the pre-*L_p *space* K_p *is a normed linear space.*

Remark 2.18 Note again that the domain of the functions x in a pre-L_p space can be altered (sometimes with necessary extra care) and that the codomain can be replaced by $\mathbb{C}$ (with obvious adaptations to the definition of the L_p norm). All such spaces are still called pre-L_p spaces and denoted, ambiguously, by K_p. In particular, the space $C([a,b],\mathbb{R})$ of continuous real-valued functions on a closed interval can be endowed with an L_p norm and thus is a K_p space.

Exercises

Exercise 2.41 Prove the generalized Pythagoras' Theorem: In an inner product space, given m pairwise orthogonal vectors $x_1, \ldots, x_m$, the equality

$$\|x_1 + \cdots + x_m\|^2 = \|x_1\|^2 + \cdots + \|x_m\|^2$$

holds.

Exercise 2.42 Prove that ℓ_2 when endowed with the operation

$$\langle x, y\rangle = \sum_{k=1}^{\infty} \overline{x_k} y_k$$

is an inner product space.

Exercise 2.43 Prove that ℓ_∞ with the ℓ_∞ norm is a normed space and that K_∞ with the L_∞ norm is a normed space.

Exercise 2.44 Let V be an inner product space. Prove the parallelogram identity

$$\|x + y\|^2 + \|x - y\|^2 = 2(\|x\|^2 + \|y\|^2)$$

for all $x, y \in V$.

Exercise 2.45 Prove that for $p \neq 2$ the space ℓ_p is not an inner product space.

Exercise 2.46 Does the equality

$$\ell_0 = \bigcap_{p>0} \ell_p$$

hold?

Exercise 2.47 For $x \in \ell_p$, with $1 \leq p < \infty$, does the equality

$$\lim_{q \to \infty} \|x\|_q = \|x\|_\infty$$

hold?

Exercise 2.48 Prove that K_2 is an inner product space.

Exercise 2.49 Let $1 \leq p \leq \infty$ and consider, for any $x \in K_p$ the sequence of samples $s(x) = (x(1), x(2), \ldots, x(k), \ldots)$. Prove that s is a linear operator $s : K_p \to \ell_p$. How do the norms $\|x\|_p$ and $\|s(x)\|_p$ compare?

Exercise 2.50 Prove Hölder's Inequality for functions and prove that K_p with the L_p norm is a normed space.

Further Reading

Introductory level texts on linear algebra typically treat linear spaces with a strong emphasis on techniques of matrices. When bases are not explicitly available, as in the infinite dimensional case, this approach must be replaced by a coordinate-free approach, as was done in this chapter. For an introductory level text treating linear algebra in a coordinate-free fashion see Chaps. 10–12 of Dummit and Foote [2]. A more advanced text with a coordinate-free approach, as well as an algebraically much deeper approach to linear algebra, making the connection between linear spaces and modules, is Roman [3]. Another text that bridges the gap between matrix-centric linear algebra and Hilbert space theory is Brown [1]. The reader seeking to enhance her intuition and ability with the Cauchy-Schwarz inequality is advised to consider Steele [4].

References

1. W.C. Brown, *A Second Course in Linear Algebra* (A Wiley-Interscience Publication, Wiley, New York, 1988) 264 p
2. D.S. Dummit, R.S. Foote, *Abstract Algebra*, 3rd edn. (Wiley, Hoboken, 2004) 932 p
3. S. Roman, *Advanced Linear Algebra*, 3rd edn., Graduate Texts in Mathematics 135 (Springer, New York, 2008) 522 p
4. J.M. Steele, *The Cauchy-Schwarz master class*, MAA Problem Books Series, Mathematical Association of America, Washington, DC (Cambridge University Press, Cambridge, 2004) 306 p

Chapter 3
Topological Spaces

Abstract This chapter, which is quite independent from the previous one, introduces topological spaces. It includes a detailed motivation for the definitions so as to assist in the digestion of the rather abstract concepts involved. The chapter is, by necessity, only a glimpse of the vast realms of topology. The presentation and the choice of concepts and results given are geared towards the applications of topology in Hilbert space theory while ensuring the reader develops a sufficient level of familiarity with the techniques (and the at times eccentric nature) of topology.

Keywords Topological space · Continuous functions · Separation axioms · Countability axioms · Hausdorff space · Compact space · Connected space · Product topology · Quotient topology · Generated topology

Topology is often described as rubber-sheet geometry, i.e., the study of geometric properties that are insensitive to stretching and shrinking, without tearing or gluing. Famously, for a topologist then, there is no difference between a cup and a donut. The importance and impact of topology on modern mathematics is quite difficult to quantify but virtually impossible to exaggerate. Without a doubt its place as one of the pillars of mathematics is secured.

The definition of a topological space is an abstraction of part of the structure of $\mathbb{R}$, specifically that part that allows one to speak of continuity and convergence. The choice for the topological notions introduced in this chapter is dictated by two needs. One need is to familiarize the reader with those topological concepts that are most relevant in the context of this book, the other is to familiarize the reader with topology proper, its main results, and its techniques. These two needs pull in slightly different directions and this chapter represents a compromise. We aim to provide the reader with sufficient motivation for the concepts while keeping things firmly grounded in analysis.

Section 3.1 presents the definition of a topology and the accompanying notion of continuity, followed by a detailed construction of the Euclidean topology on the real numbers $\mathbb{R}$. The associated notion of continuity is shown to be equivalent to the usual one, motivating the definitions. Section 3.2 is a basic study of convergence in the topological setting, its relation to cluster points, and its behaviour under the first countability axiom. Section 3.3 is concerned with techniques for constructing topologies, in particular by forcing a collection of sets to be open sets and by forcing a

© Springer International Publishing Switzerland 2015

C. Alabiso and I. Weiss, *A Primer on Hilbert Space Theory*,
UNITEXT for Physics, DOI 10.1007/978-3-319-03713-4_3

collection of functions to be continuous. Coproducts, products, and quotients are then given as particular examples. Section 3.4 discusses the Hausdorff separation property and topological connectivity, and Sect. 3.5 is a study of compactness, developing enough material to prove the Heine-Borel Theorem from a topological perspective.

3.1 Topology—Definition and Elementary Results

The definition of a topology is quite simple; a certain collection of subsets of a set satisfying some simple to state axioms. However, without further motivation, the axioms may appear arbitrary and not too palatable. To assist the reader digest the notion, the definitions below are immediately succeeded by a detailed motivating discussion taking place in the familiar setting of the real numbers $\mathbb{R}$. The motivation we give is of a topology as the result of stripping away redundancies in the definition of continuity. We then explore various examples of topological spaces.

3.1.1 Definition and Motivation

We present the definition of a topological space followed by the definition of a continuous function between topological spaces. The abstract definitions are then exemplified in the context of the real numbers, establishing the Euclidean topology on $\mathbb{R}$, and, through the process, justifying the abstract definitions.

Definition 3.1 A *topology* τ_X on an arbitrary set X is a collection of subsets of X such that the following conditions hold.

1. Both the empty set $\emptyset$ and the entire set X are members of τ_X.
2. τ_X is stable under arbitrary unions. That is, if $\{U_i\}_{i \in I}$ is a family of elements in τ_X, then

$$\bigcup_{i \in I} U_i \in \tau_X.$$

3. τ_X is stable under finite intersections. That is, if $U_1, \ldots, U_m$ are members of τ_X, then

$$U_1 \cap \cdots \cap U_m \in \tau_X.$$

The pair (X, τ_X) is called a *topological space* and the members of τ_X are referred to as *open* sets. Often though, we speak of the topological space X without explicitly mentioning τ_X. We write (X, τ) if there is no need to syntactically remind ourselves that the collection τ consists of the open sets of X.

Remark 3.1 Clearly, stability under finite intersections is equivalent to the condition that $U_1 \cap U_2 \in \tau_X$ for any *two* open sets $U_1, U_2 \in \tau_X$.

For the following definition, and as a general advise for safely taking the first few steps into the realm of topology, the reader may wish to review the basic properties of inverse images (e.g. Sect. 1.2.9 of the Preliminaries).

Definition 3.2 Let (X, τ_X) and (Y, τ_Y) be topological spaces. A function $f : X \to Y$ is said to be *continuous* if $f^{-1}(U)$ is an open set in X whenever U is an open set in Y. In other words, f is continuous if

$$U \in \tau_Y \implies f^{-1}(U) \in \tau_X.$$

The following results establish a particular topology on the set $\mathbb{R}$ of real numbers, and investigate the notion of continuity with respect to that topology. The aim is two-fold; to exemplify the definitions in a concrete and familiar setting, and to motivate the definitions.

Recall that a function $f : \mathbb{R} \to \mathbb{R}$ is said to be continuous if, intuitively, small changes in the value of x cause small changes in the value of $f(x)$. Formally, f is continuous if for every point $x_0 \in \mathbb{R}$ and $\varepsilon > 0$ one can find a $\delta > 0$ such that if $|x - x_0| < \delta$, then $|f(x) - f(x_0)| < \varepsilon$. The absolute value function is used here in the role of a measurement device for distances. However, the notion of continuity is quite insensitive to numerous changes one can perform on the absolute value function. Intuitively, continuity is a local phenomenon, and so if one truncates the absolute value function and defines

$$|x|_t = \begin{cases} |x| & \text{if } |x| < 1 \\ 1 & \text{otherwise,} \end{cases}$$

then continuity with respect to $|x|$ and with respect to $|x|_t$ coincide. Scaling effects, such as defining $|x|_2 = 2|x|$, also have no effect on continuity in the precise sense that continuity with respect to $|x|$ and with respect to $|x|_2$ coincide.

The decision to use the absolute value function in formulating the definition of continuity (or, more generally, that of limit) is thus revealed to be an arbitrary choice among many equivalently valid alternatives. It is now a natural question whether one can distill a definition of continuity devoid of any arbitrary and irrelevant structure and consequently furnish a better understanding of the concept of continuity. The affirmative answer is given as follows. Call a subset $U \subseteq \mathbb{R}$ *open* if for every $x \in U$ there exists an $\varepsilon > 0$ such that $B_\varepsilon(x) \subseteq U$, where we define

$$B_\varepsilon(x) = \{y \in \mathbb{R} \mid |x - y| < \varepsilon\} = (x - \varepsilon, x + \varepsilon),$$

the *open interval* of length ε about x. With the aim of utilizing these sets to define a topology on $\mathbb{R}$, we first prove that any open interval is an open set.

Proposition 3.1 *The open interval $B_\varepsilon(x) = (x - \varepsilon, x + \varepsilon)$, for all $x \in \mathbb{R}$ and $\varepsilon > 0$, is an open set.*

Proof Given $y \in B_\varepsilon(x) = (x - \varepsilon, x + \varepsilon)$, we are required to find a $\delta > 0$ such that $B_\delta(y) = (y - \delta, y + \delta) \subseteq B_\varepsilon(x)$. Let

$$\delta = \varepsilon - |x - y|$$

and note that $\delta > 0$. It is now elementary algebra to verify that this δ suits our purposes. □

We can now present our first example of a topology; an important enough example that we state it as a theorem.

Theorem 3.1 *The collection τ of all open sets $U \subseteq \mathbb{R}$ is a topology on $\mathbb{R}$.*

Proof The empty set vacuously satisfies the condition for being an open set, simply since it has no points at all. The set $\mathbb{R}$ itself is clearly an open set, since if $x \in \mathbb{R}$, then $x \in B_1(x) \subseteq \mathbb{R}$. Thus the first condition in the definition of a topology is satisfied. To show stability under arbitrary unions, suppose that $\{U_i\}_{i \in I}$ is a family of open sets in $\mathbb{R}$ and we need to show that

$$U = \bigcup_{i \in I} U_i$$

is open. Indeed, if $x \in U$, then $x \in U_i$ for some $i \in I$. But since U_i is open, there exists a $\delta > 0$ with

$$B_\delta(x) \subseteq U_i \subseteq U,$$

as required. It remains to show for any two open sets $U_1, U_2 \subseteq \mathbb{R}$ that $U_1 \cap U_2$ is open. Indeed, if $x \in U_1 \cap U_2$, then there are $\delta_1 > 0$ and $\delta_2 > 0$ with

$$B_{\delta_1}(x) \subseteq U_1$$
$$B_{\delta_2}(x) \subseteq U_2.$$

Setting $\delta = \min\{\delta_1, \delta_2\}$ (and noting that $\delta > 0$), one easily sees that

$$B_\delta(x) \subseteq B_{\delta_1}(x) \cap B_{\delta_2}(x) \subseteq U_1 \cap U_2,$$

as required, and thus completing the proof. □

Definition 3.3 The collection of all open sets $U \subseteq \mathbb{R}$ as just defined is called the *Euclidean topology* on $\mathbb{R}$. It is also commonly referred to as the *standard* topology or the *ordinary* topology.

The open sets turn out to contain enough information about the space $\mathbb{R}$ in order to completely characterize continuity, as follows.

Theorem 3.2 *A function $f : \mathbb{R} \to \mathbb{R}$ is continuous in the usual meaning if, and only if, $f^{-1}(U)$ is an open set for every open set $U \subseteq \mathbb{R}$.*

Proof Suppose f is continuous, and let U be an open set. Our aim is to show that $f^{-1}(U)$ is open. Let $x_0 \in f^{-1}(U)$, that is $y_0 = f(x_0) \in U$, and our aim is to find a $\delta > 0$ such that

$$B_\delta(x_0) \subseteq f^{-1}(U).$$

Since U is open, there exists an $\varepsilon > 0$ such that

$$B_\varepsilon(y_0) = (y_0 - \varepsilon, y_0 + \varepsilon) \subseteq U.$$

By continuity of f, there exists a $\delta > 0$ such that $|f(x) - y_0| < \varepsilon$, provided that $|x - x_0| < \delta$. But that means that for every $x \in B_\delta(x_0)$

$$f(x) \in B_\varepsilon(y_0) \subseteq U.$$

In other words, $x \in f^{-1}(U)$ for all x in the set $B_\delta(x_0)$, and so

$$B_\delta(x_0) \subseteq f^{-1}(U),$$

which is what we set out to obtain.

In the other direction, suppose that $f^{-1}(U)$ is an open set whenever U is an open set. Given a point $x_0 \in \mathbb{R}$ and an $\varepsilon > 0$, let $y_0 = f(x_0)$ and consider the set $U = B_\varepsilon(y_0)$, and notice that U is itself an open set. By assumption, the set $f^{-1}(U)$ is an open set, and it contains x_0. There is thus a $\delta > 0$ such that $B_\delta(x_0) \subseteq f^{-1}(U)$. Thus, if $|x - x_0| < \delta$, then $x \in B_\delta(x_0)$, and therefore $f(x) \in B_\varepsilon(y_0)$, showing that $|f(x) - f(x_0)| < \varepsilon$. In other words, f is continuous at x_0. Since x_0 was arbitrary, f is continuous. □

Recalling Definition 3.2, the result above can be restated as follows.

Theorem 3.3 *A function $f : \mathbb{R} \to \mathbb{R}$ is continuous in the usual sense if, and only if, it is continuous with respect to the Euclidean topology on $\mathbb{R}$.*

This result justifies and motivates Definition 3.2 and, a-priori, the definition of a topology. Let us pause to reflect on the situation right now. The familiar notion of continuity for functions $f : \mathbb{R} \to \mathbb{R}$ was seen to be faithfully captured by the collection of open sets $U \subseteq \mathbb{R}$. The collection of open sets was defined in terms of $B_\varepsilon(x) = \{y \in \mathbb{R} \mid |x - y| < \varepsilon\}$, thus still directly using the absolute value function. It would thus seem that we had accomplished nothing except for hiding the absolute value function in some fancy notation. However, something quite substantial was achieved. Consider the scaled absolute value function $|x|_2 = 2 \cdot |x|$, and let us define

$$B_{2,\varepsilon}(x) = \{y \in \mathbb{R} \mid |x - y|_2 < \varepsilon\} = (x - \frac{\varepsilon}{2}, x + \frac{\varepsilon}{2})$$

for all $x \in \mathbb{R}$ and $\varepsilon > 0$. Let us further say that a set $U \subseteq \mathbb{R}$ is 2-open if for all $x \in U$ there exists an $\varepsilon > 0$ such that $B_{2,\varepsilon}(x) \subseteq U$. In other words, we repeat the definition of open sets, replacing the absolute value function by a scaled version of it.

Proposition 3.2 *The collection of open sets and the collection of 2-open sets are the same.*

Proof Notice that

$$B_{\varepsilon/4}(x) \subseteq B_{2,\varepsilon}(x) \subseteq B_\varepsilon(x)$$

holds for all $x \in \mathbb{R}$ and $\varepsilon > 0$. Thus, if U is an open set and $x \in U$, then

$$B_\varepsilon(x) \subseteq U$$

for some $\varepsilon > 0$, but then

$$B_{2,\varepsilon}(x) \subseteq B_\varepsilon(x) \subseteq U$$

and so U is also 2-open. Conversely, if U is 2-open and $x \in U$, then

$$B_{2,\varepsilon}(x) \subseteq U$$

for some $\varepsilon > 0$, but then

$$B_{\varepsilon/4}(x) \subseteq B_{2,\varepsilon}(x) \subseteq U$$

and so U is also open. □

Similarly, one may show that the collection of open sets is insensitive to other changes in the absolute value function, such as scaling by other factors or various truncations. It is not a coincidence that those changes to the absolute value function that are immaterial to the notion of continuity are also immaterial to the concept of open set. Indeed, Theorem 3.2 characterizes continuity in terms of the open sets alone.

Thus, by concentrating on the open sets one is freed from the irrelevant (for the purposes of continuity) particularities of the absolute value function. The definition of a topology is thus an abstraction of a particular formulation of continuity which avoids directly mentioning any particular concept of $\mathbb{R}$ which is irrelevant to the notion of continuity. As with any abstraction, once it is made, a plethora of new situations emerges, situations that superficially have little or nothing to do with the geometric notion of continuity. Introducing a topology on a set immediately allows one to import a significant amount of geometric intuition, machinery, and analogy into an a-priori non-geometric context.

3.1.2 More Examples

We present more topological spaces, illustrating the versatility of the formalism. We remark that there are far more interesting examples of topological spaces than can be recounted in this book. The choice of the examples given below is dictated by the focus of the book; the examples are chosen to elucidate the forthcoming topological concepts, while avoiding delving into the intricacies of the more obscure examples of topological spaces, thus allowing the reader to ease into the topological framework while remaining firmly grounded in analysis.

Example 3.1 (*Discrete And Indiscrete Topologies*) Any set X supports two extreme topologies on it. One is the collection $\{\emptyset, X\}$, called the *indiscrete topology* on X and the other is the collection of all subsets of X, called the *discrete topology* on X. It is immediate to verify that these are indeed topologies, and that they are distinct topologies except when $|X| \leq 1$.

Example 3.2 (*The Sierpinski Space*) Let $\mathbb{S} = \{0, 1\}$ be a set with two elements. The collection

$$\{\emptyset, \mathbb{S}, \{1\}\}$$

is readily seen to be a topology on $\mathbb{S}$ giving rise to what is known as the *Sierpinski* space.

Example 3.3 Let $X = \{a, b, c, d, e\}$ be a nonempty set consisting of five elements. Of the collections

$$\begin{aligned}
\tau &= \{X, \emptyset, \{a\}, \{c, d\}, \{a, c, d\}, \{b, c, d, e\}\} \\
G &= \{X, \emptyset, \{a\}, \{c, d\}, \{a, c, d\}, \{b, c, d\}\} \\
H &= \{X, \emptyset, \{a\}, \{c, d\}, \{a, c, d\}, \{a, b, d, e\}\}
\end{aligned}$$

τ is a topology on X as the axioms are easily confirmed by inspection. G is not a topology on X, since $\{a\} \cup \{b, c, d\} = \{a, b, c, d\} \notin G$. The collection H is also not a topology on X, since $\{c, d\} \cap \{a, b, d, e\} = \{d\} \notin H$.

Example 3.4 (*The Cofinite and Cocountable Topologies*) Let X be a set and τ the family of all subsets U of X whose complement $X - U$ is finite, and if needed, manually include $\emptyset$ in τ. We show that τ is a topology on X. First, $\emptyset \in \tau$ by definition, and since $X - X = \emptyset$ is finite, it follows that $X \in \tau$, and so the first condition in the definition of topology is satisfied. Stability under arbitrary unions is immediate, so finally consider two sets U_1, U_2 with finite complements in X. Since

$$X - (U_1 \cap U_2) = (X - U_1) \cup (X - U_2)$$

and since both $(X - U_1)$ and $(X - U_2)$ are finite, it follows that $U_1 \cap U_2$ has finite complement too. This topology is called the *cofinite topology* on X. Notice that if

X is finite, then the cofinite topology coincides with the discrete topology on X (Example 3.1).

Similarly, call a subset $U \subseteq X$ *cocountable* if its complement $X - U$ is a countable set. The collection of all cocountable subsets of X (with $\emptyset$ added in manually) also forms a topology on X called the *cocountable topology*. The proof is similar to the proof given above, with the necessary care when handling countable sets (see Sect. 1.2.13 of the Preliminaries).

3.1.3 Elementary Observations

We explore some immediate properties of topological spaces and of continuous functions, and we discuss the notion of homeomorphism. We start off with a useful criterion for detecting open sets in any topological space.

Proposition 3.3 *A subset V of a topological space X is open if, and only if, for all $x \in V$ there exists an open set U_x such that*

$$x \in U_x \subseteq V.$$

Proof If V is open, then clearly for every $x \in V$ one may choose $U_x = V$. In the other direction, the stated condition implies that

$$V = \bigcup_{x \in V} U_x$$

and so V is expressed as the union of open sets, and is thus itself open. □

Proposition 3.4 *The composition $g \circ f : X \to Z$ of any two continuous functions $X \xrightarrow{f} Y \xrightarrow{g} Z$ between topological spaces is itself continuous.*

Proof Given an open set U in Z, we need to verify that $(g \circ f)^{-1}(U)$ is open in X. Indeed,

$$(g \circ f)^{-1}(U) = f^{-1}(g^{-1}(U))$$

and since g is continuous it follows that $g^{-1}(U)$ is open in Y, and since f, too, is continuous it follows that $f^{-1}(g^{-1}(U))$ is open in X. □

Proposition 3.5 *Given two topologies, τ_1 and τ_2, on the same set X, the identity function* $\mathrm{id} : (X, \tau_1) \to (X, \tau_2)$ *is continuous if, and only if, $\tau_2 \subseteq \tau_1$.*

Proof Notice that for the identity function $\mathrm{id} : X \to X$ and any subset $U \subseteq X$ (open or not) one has $\mathrm{id}^{-1}(U) = U$. Thus, the condition for continuity is precisely that $U \in \tau_1$ for all $U \in \tau_2$, in other words, that $\tau_2 \subseteq \tau_1$. □

Definition 3.4 Given two topologies τ_1 and τ_2 on the same set X, if $\tau_1 \subseteq \tau_2$, then τ_1 is said to be *coarser* than τ_2 while τ_2 is said to be *finer* than τ_1.

Notice that the discrete topology on X is the finest topology among all topologies on X while the indiscrete topology is the coarsest one. In general, different topologies on the same set may be incomparable, i.e. one need not be contained in the other.

Definition 3.5 A function $f : X \to Y$ between topological spaces is said to be a *homeomorphism* if f is bijective and both $f : X \to Y$ and $f^{-1} : Y \to X$ are continuous functions. If there exists a homeomorphism between X and Y, then X and Y are said to be *homeomorphic* spaces.

Remark 3.2 One should not confuse between the terms homeomorphism and homomorphism. The term homomorphism in algebra is typically used to indicate a structure preserving function, with isomorphism used for the invertible structure preserving functions. In topology, the structure preserving functions are precisely the continuous functions, and the invertible structure preserving functions are called homeomorphisms.

Note that $f : X \to Y$ is a homeomorphism if, and only if, for all $U \subseteq Y$

$$f^{-1}(U) \in \tau_X \iff U \in \tau_Y.$$

Homeomorphic spaces thus have essentially the same collections of open sets, up to a renaming of the elements. Intuitively, two spaces are homeomorphic if one can be continuously changed, without gluing or tearing, to obtain the other. While homeomorphic spaces must have the same cardinality (since a homeomorphism is in particular a bijection), the converse quite often fails since both the bijection and its inverse must be continuous. For instance, the circumference of a circle and a line segment (say in $\mathbb{R}^2$) have the same cardinality, but they are not homeomorphic. This is intuitively quite clear since (it would appear that) the only way to create the hole that the circle encloses is to glue the two ends of the line segment, but that changes the topology (i.e. glueing is a continuous operation but its inverse, tearing, is not). However, turning such intuitive arguments into formal ones is usually not straightforward.

Generally speaking, it can be quite tricky to prove that two topological spaces are not homeomorphic. A common technique in establishing such a result is the following one. A property P about topological spaces is said to be a *topological invariant* if for all homeomorphic topological spaces X and Y, X has property P precisely when Y does. To show that two spaces are not homeomorphic, it thus suffices to find a topological invariant that only one of the two spaces possesses. The topological concepts introduced in the rest of this chapter often provide one with a suitable topological invariant.

3.1.4 Closed Sets

With every subset S of a set X one may associate the complementary set $S^c = X - S$, giving rise to a bijective operation $\mathscr{P}(X) \to \mathscr{P}(X)$ on the set $\mathscr{P}(X)$ of all subsets of X. It is thus clear that any concept given in terms of subsets of X gives rise, by taking complements, to what must essentially be an equivalent concept. This general observation holds for the definition of a topology, and we now devote some time to the relevant details.

We first observe that the Sierpinski space $\mathbb{S} = \{0, 1\}$ from Example 3.2 can be used to identify the open sets in an arbitrary topological space.

Proposition 3.6 *There exists a bijective correspondence between the open sets of a topological space X and continuous functions $f : X \to \mathbb{S}$.*

Proof Recall (e.g. Sect. 1.2.10 of the Preliminaries) that with every subset $S \subseteq X$ one may associate the indicator function

$$f_S(x) = \begin{cases} 1 & \text{if } x \in S \\ 0 & \text{if } x \notin S \end{cases}$$

and that this correspondence is a bijection between all subsets of X and all functions $X \to \mathbb{S}$. Now, for $f_S : X \to \mathbb{S}$ to be continuous, the inverse image of every open set in $\mathbb{S}$, namely the sets $\emptyset$, $\mathbb{S}$, and $\{1\}$, must be open in X. The set $f_S^{-1}(\emptyset) = \emptyset$ is always open in X and similarly so is $f_S^{-1}(\mathbb{S}) = X$. So, the only further condition imposed by continuity is that $f_S^{-1}(\{1\})$ must be open too, and since the latter set is precisely S, we see that continuous functions $f : X \to \mathbb{S}$ correspond bijectively to the open subsets of X. □

Clearly, every function $f : X \to \mathbb{S}$ is equally well completely determined by the inverse image of 0 instead of the inverse image of 1. We thus see that there is also a correspondence between continuous functions $f : X \to \mathbb{S}$ and the *complements of the open sets* in X, giving rise to the following definition.

Definition 3.6 A subset F of a topological space X is said to be *closed* provided its complement $X - F$ is open.

The preceding discussion implies that a topology can be specified by the collection of closed sets. The details are given in the following result.

Theorem 3.4 *Let X be an arbitrary set. A given collection $\mathscr{F}$ of subsets of X is the collection of closed sets for some topology on X if, and only if, the following conditions hold.*

1. *Both the empty set $\emptyset$ and the entire set X are members of $\mathscr{F}$.*
2. *$\mathscr{F}$ is stable under finite unions. That is, if $F_1, \ldots, F_m$ are members of $\mathscr{F}$, then*

$$F_1 \cup \cdots \cup F_m \in \mathscr{F}.$$

3. *$\mathscr{F}$ is stable under arbitrary intersections. That is, if $\{F_i\}_{i\in I}$ are members of $\mathscr{F}$, then*

$$\bigcap_{i\in I} F_i \in \mathscr{F}.$$

Moreover, a collection $\mathscr{F}$ satisfying the conditions above is the collection of closed sets for a unique topology on X.

Proof Applying De Morgan's Laws (see Sect. 1.2.5 of the Preliminaries), the details are straightforward and are left to the reader. We only stress out here that the collection

$$\{X - F \mid F \in \mathscr{F}\}.$$

is the unique topology determined by $\mathscr{F}$. □

Example 3.5 If a set X is given the discrete topology, then every subset of it is closed, while if X is given the indiscrete topology, then the only closed subsets of X are $\emptyset$ and X itself. In the Sierpinski space $\mathbb{S} = \{0, 1\}$, the closed sets are: $\emptyset$, $\mathbb{S}$, $\{0\}$.

Example 3.6 Consider $\mathbb{R}$ with the Euclidean topology (Definition 3.3). It is easy to verify that for $a, b \in \mathbb{R}$ with $a < b$, the open interval (a, b) is an open set and the closed interval $[a, b]$ is a closed set. The open rays (a, ∞) and $(-\infty, a)$ are also open sets, and the closed rays $[a, \infty)$ and $(-\infty, a]$ are closed sets. Every singleton set $\{a\}$ is a closed set. Consequently, every finite set $F \subseteq \mathbb{R}$ is closed.

Remark 3.3 It should be emphasized that in general an arbitrary union of closed sets need not be closed nor does an arbitrary intersection of open sets need be open. For instance, consider $\mathbb{R}$ with the Euclidean topology (Definition 3.3). Since every singleton set $\{x\}$ is closed, every subset $X \subseteq \mathbb{R}$ is the union of closed sets, i.e.

$$X = \bigcup_{x\in X} \{x\}$$

but, of course, not every subset of $\mathbb{R}$ is closed. Similarly, for any $x \in \mathbb{R}$, one has

$$\{x\} = \bigcap_{\varepsilon>0} (x - \varepsilon, x + \varepsilon)$$

showing that $\{x\}$ is an intersection of open sets, but is itself not open.

Remark 3.4 It is important to realize that a set is not a door. A set need not be open nor closed, while it may be both open and closed. For instance, consider the topology τ from Example 3.3. One easily verifies that $\{a, c\}$ is not closed nor open and that $\{a\}$ is both closed and open. Sets that are both closed and open are said to be *clopen*. Notice that the empty set $\emptyset$ and X itself are clopen in any topological space. If these are the only clopen sets in X, than X is said to be *connected*, a property we explore further in Sect. 3.4.

The dual nature of open and closed sets is seen in the following result. Recall that a function $f : X \to Y$ between topological spaces is continuous if $f^{-1}(U)$ is open in X for every open set U in X.

Theorem 3.5 *A function $f : X \to Y$ between topological spaces is continuous if, and only if, $f^{-1}(F)$ is a closed set in X for every closed set F in Y.*

The proof is, as it should be, a tautology that the reader is urged to clarify for herself.

The duality between open and closed sets is quite helpful since while dual to each other, open and closed sets have different topological qualities and thus it may be easier to establish a result in terms of closed sets rather than open ones, or vice versa. The reader should be warned though that not all concepts that may appear dual are in fact dual.

Definition 3.7 A function $f : X \to Y$ between topological spaces is

- an *open mapping* if $f(U)$ is open in Y for every open set U in X.
- a *closed mapping* if $f(F)$ is closed in Y for every closed set F in X.

The reader is invited to find examples of open mappings that are not closed as well as closed ones that are not open.

3.1.5 Bases and Subbases

Generally speaking, a topology is a very large collection of subsets. For instance, the Euclidean topology on $\mathbb{R}$ has uncountably many open sets. It is thus often desirable to obtain smaller (or at least more manageable) collections that give one access to the entire topology. This is what bases and subbases are designed to achieve.

Definition 3.8 A collection $\mathscr{B}$ of open sets in a topological space X is a *basis* (or a *base*) for the topology τ_X if for every open set $U \subseteq X$ and $x \in U$, there exists an element $B \in \mathscr{B}$ such that

$$x \in B \subseteq U.$$

The elements of $\mathscr{B}$ are called *basis elements.*

Example 3.7 For every topological space (X, τ_X), the collection $\mathscr{B} = \tau_X$ is, quite trivially, a basis. More interestingly, the collection $\{B_\varepsilon(x) \mid x \in \mathbb{R},\ \varepsilon > 0\}$ is a basis for the Euclidean topology on $\mathbb{R}$. Indeed, if $U \subseteq \mathbb{R}$ is open, then, by definition, for each $x \in U$ there is an $\varepsilon > 0$ such that

$$x \in B_\varepsilon(x) \subseteq U.$$

If X is an arbitrary set endowed with the indiscrete topology, then the collection $\{X\}$ is a basis. Indeed, there are at most two open sets in the indiscrete topology, namely $\emptyset$ and X. So if U is open and $x \in U$, then necessarily $U = X$ and clearly

$$x \in X \subseteq X.$$

If X is endowed with the discrete topology, then the collection $\{\{x\} \mid x \in X\}$ of all singleton subsets of X is a basis. Indeed, in the discrete topology every $U \subseteq X$ is open, and if $x \in U$, then clearly

$$x \in \{x\} \subseteq U.$$

Definition 3.9 A collection $\mathscr{A}$ of open sets in a topological space X is a *subbasis* for the topology if the collection $\{A_1 \cap \cdots \cap A_m \mid A_1, \ldots, A_m \in \mathscr{A},\ m \geq 0\}$ of all finite intersections of elements of $\mathscr{A}$ is a basis for the topology.

Remark 3.5 By convention, the empty intersection (i.e. the case $m = 0$ above) is interpreted to be the set X itself.

Example 3.8 Consider the set of real numbers $\mathbb{R}$ with the Euclidean topology. The collection

$$\{(a, \infty) \mid a \in \mathbb{R}\} \cup \{(-\infty, b) \mid b \in \mathbb{R}\}$$

of all open rays is a subbasis. Indeed, the open rays are open sets and since

$$B_\varepsilon(a) = (a - \varepsilon, \infty) \cap (-\infty, a + \varepsilon)$$

we see that the collection of all finite intersections of the open rays contains the basis we presented above. It follows that the collection is a basis since, in general, if $\mathscr{B}' \supseteq \mathscr{B}$ are collections of open sets and $\mathscr{B}$ is a basis, then so is $\mathscr{B}'$.

We now show that bases can be used directly to detect continuity of functions.

Proposition 3.7 *Let $f : X \to Y$ be a function between topological spaces and let $\mathscr{B}$ be a basis for the topology on Y. Then f is continuous if, and only if, $f^{-1}(B)$ is open in X for all basis elements $B \in \mathscr{B}$.*

Proof If f is continuous, then $f^{-1}(U)$ is open for all open sets $U \subseteq Y$. Since any $B \in \mathscr{B}$ is in particular an open set in Y, it follows that $f^{-1}(B)$ is open in X. In the other direction, assume that $f^{-1}(B)$ is open for all $B \in \mathscr{B}$ and let U be an arbitrary open set in Y. It follows at once from the definition of basis that we may express U as a union of basis elements

$$U = \bigcup_{x \in U} B_x.$$

It then holds that

$$f^{-1}(U) = \bigcup_{x \in U} f^{-1}(B_x)$$

and since each $f^{-1}(B_x)$ is assumed open, the set $f^{-1}(U)$ is thus expressed as the union of open sets, and so is open. Thus $f^{-1}(U)$ is open for all open sets U in Y, and thus f is continuous. □

We close this section by introducing a local variant of the notion of basis.

Definition 3.10 Let X be a topological space and $x \in X$. A collection $\mathscr{B}$ of open subsets of X is said to be a *local basis* at x if for all $B \in \mathscr{B}$

$$x \in B$$

and if for every open set U in X with $x \in U$, there is an element $B_x \in \mathscr{B}$ (called a *basis element)* with

$$B_x \subseteq U.$$

Every basis $\mathscr{B}$ for the topology on X clearly gives rise to the local basis

$$\mathscr{B}_x = \{B \in \mathscr{B} \mid x \in B\}$$

at x. It is also the case that if for every $x \in X$ one has a local basis $\mathscr{B}_x$ at x, then the collection

$$\mathscr{B} = \bigcup_{x \in X} \mathscr{B}_x$$

is a basis for the topology on X. We leave the verification of these simple claims to the reader.

Exercises

Exercise 3.1 How many topologies are there on a set with three elements?

Exercise 3.2 Characterize all continuous functions $f : \mathbb{S} \to \mathbb{R}$ from the Sierpinski space to $\mathbb{R}$ with the Euclidean topology.

Exercise 3.3 Prove Theorem 3.5.

Exercise 3.4 Show that the concepts of open function, of closed function, and of continuous function are independent of each other.

Exercise 3.5 Give an example of a topological space with infinitely many open sets in which any two non-empty open sets have non-empty intersection.

Exercise 3.6 Let X and Y be sets. When X is equipped with the discrete topology and Y with an arbitrary topology, show that every function $f : X \to Y$ is continuous.

Exercise 3.7 Let X and Y be sets. When Y is equipped with the indiscrete topology and X with an arbitrary topology, show that every function $f : X \to Y$ is continuous.

Exercise 3.8 Show that any constant function $f : X \to Y$ between topological spaces is continuous.

Exercise 3.9 Show that any continuous function $f : X \to Y$, when Y is given the discrete topology and X is given the indiscrete topology, is a constant function.

Exercise 3.10 Let $X = \mathbb{R}$ endowed with the Euclidean topology and let $Y = \mathbb{R}$ with the cofinite topology. Is the identity function $\mathrm{id} : X \to Y$ continuous? Is the identity function $\mathrm{id} : Y \to X$ continuous?

3.2 Subspaces, Point-Set Relationships, and Countability Axioms

This section introduces geometric notions regarding various possible relationships between a point and a set in a topological space. We also consider countability axioms and study the consequences they entail for some of these notions.

3.2.1 Subspaces and Point-Set Relationships

Any subset of a topological space naturally inherits a topology and thus one may speak of subspaces, which we discuss first. Next, various qualitative aspects of a given point and a subset of a topological space are introduced and some basic results are established.

Definition 3.11 (*Subspace Topology*) Let (X, τ_X) be a topological space and $Y \subseteq X$ a subset. The collection

$$\tau_Y = \{U \cap Y \mid U \in \tau_X\}$$

is (easily seen to be) a topology on Y called the *subspace* topology. The topological space (Y, τ_Y) is called a *topological subspace* of (X, τ_X), or simply a *subspace* if the topological context is clear.

Proposition 3.8 *The subspace topology on a subset Y of a topological space X is the smallest topology on Y making the inclusion function $i : Y \to X$ continuous.*

Proof Noticing that

$$i^{-1}(U) = Y \cap U$$

for all $U \subseteq X$ one sees that the inclusion function $i : Y \to X$ is continuous precisely if $Y \cap U$ is open in Y whenever $U \subseteq X$ is open in X. This condition clearly holds for the subspace topology τ_Y on Y. Further, if τ is any topology on Y, then the condition that $i : Y \to X$ is continuous with respect to the topologies τ on Y and τ_X on X amounts to the condition

$$U \in \tau_X \implies U \cap Y \in \tau,$$

and thus that $\tau_Y \subseteq \tau$, proving the minimality claim of the subspace topology τ_Y. □

Given a subset and a point in a topological space, the open sets may be used to identify the qualitative relative location of the point in relation to the set. This is the subject of the following definition.

Definition 3.12 Let X be a topological space, $S \subseteq X$ a subset, and $x \in X$ a point.

1. x is an *interior* point of S if there exists an open set U such that $x \in U \subseteq S$.
2. x is an *exterior* point of S if there exists an open set U such that $x \in U \subseteq X - S$.
3. x is a *boundary* point of S if every open set U that contains x, contains elements in S as well as elements in $X - S$.
4. x is a *cluster* point or an *accumulation* point of S, if every open set U which contains x, contains elements of $S - \{x\}$.
5. x is an *isolated* point of S, if $\{x\}$ is an open set.

Example 3.9 Let us consider the space $\mathbb{R}$ with its Euclidean topology. It is easily seen that for $S = [a, b]$, a closed interval, as well as $S = (a, b)$, an open interval:

1. A point x is interior precisely when $a < x < b$.
2. A point x is exterior precisely when either $x < a$ or $x > b$.
3. A point x is a boundary point precisely when $x = a$ or $x = b$.
4. A point x is a cluster point precisely when $a \leq x \leq b$.
5. S has no isolated points, but the space $S = [0, 1] \cup \{2\}$, as a subspace of $\mathbb{R}$ with the Euclidean topology, admits the point $x = 2$ as its only isolated point.

Further, in $\mathbb{R}$ with the Euclidean topology, it is easy to see that every real number x is a cluster point of the subset $\mathbb{Q}$ of rational numbers. In general, an isolated point x is never a cluster point of any set S, simply since for the open set $U = \{x\}$ one always has $U \cap (S - \{x\}) = \emptyset$.

Definition 3.13 Let X be a topological space and $S \subseteq X$ a subset.

1. The *interior* of S is the set $\text{int}(S) = \{x \in X \mid x \text{ is an interior point of } S\}$.
2. The *exterior* of S is the set $\text{ext}(S) = \{x \in X \mid x \text{ is an exterior point of } S\}$.
3. The *boundary* of S is the set $\partial(S) = \{x \in X \mid x \text{ is a boundary point of } S\}$.
4. The *derived set* of S is the set $S' = \{x \in X \mid x \text{ is a cluster point of } S\}$.
5. The *closure* of S is the set $\overline{S} = S \cup S'$.

Proposition 3.9 *Let X be a topological space and $S \subseteq X$ a subset. It then holds that*

1. *The interior* $\text{int}(S)$ *is the union of all open sets contained in S.*
2. *The exterior* $\text{ext}(S)$ *is the union of all open sets disjoint from S.*
3. *The closure $\overline{S}$ is the intersection of all closed sets that contain S.*
4. $\overline{S} = S \cup \partial(S)$.

Proof

1. Let $\mathscr{U}$ be the set of all open sets $U \subseteq X$ with $U \subseteq S$, and let

$$W = \bigcup_{U \in \mathscr{U}} U.$$

We need to show that $W = \text{int}(S)$. If $x \in W$, then $X \in U$ for some $U \in \mathscr{U}$, and since $x \in U \subseteq S$ we may conclude that $x \in \text{int}(S)$. Conversely, if $x \in \text{int}(S)$, then there exists an open set U such that $x \in U \subseteq S$, and thus $x \in W$.
2. Notice that $\text{ext}(S) = \text{int}(X - S)$, and now apply 1.
3. Let $\mathscr{F}$ be the set of all closed sets $F \subseteq X$ with $S \subseteq F$, and let

$$G = \bigcap_{F \in \mathscr{F}} F.$$

We need to show that $G = \overline{S}$. Let $x \in G$, and thus $x \in F$ for all $F \in \mathscr{F}$, and our aim is to show that $x \in \overline{S}$. If $x \in S$, then $x \in S \cup S' = \overline{S}$ so we may proceed under the assumption that $x \notin S$. We will show that in this case $x \in S'$. Indeed, let $U \subseteq X$ be an open set with $x \in U$ and suppose that $U \cap (S - \{x\}) = \emptyset$, which simplifies to $U \cap S = \emptyset$ since $x \notin S$. But then the closed set $F = X - U$ contains S, and thus $x \in F$, contradicting the fact that $x \in U$. In the other direction, suppose that $x \in \overline{S} = S \cup S'$, and we aim to show that $x \in G$, namely that $x \in F$ for every closed set F with $S \subseteq F$. If $x \in S$, then clearly $x \in F$, and we may thus proceed under the assumption that $x \notin S$, and thus that $x \in S'$. Consider the open set $U = X - F$ and note that $U \cap (S - \{x\}) = U \cap S = \emptyset$. If $x \notin F$, then $x \in U$ and we obtain a contradiction with $x \in S'$. It follows that $x \in F$, as needed.
4. To show that $\overline{S} \subseteq S \cup \partial S$, suppose that $x \in \overline{S}$ but $x \notin S$. Given an open set U with $x \in U$, it follows that $U \cap S \neq \emptyset$. Since $x \notin S$, namely $x \in X - S$, we also have that $x \in U \cap (X - S)$. In other words, any open set containing x, contains points from S as well as from $X - S$, thus x is a boundary point of S. In the other direction, suppose that $x \in S \cup \partial S$, and we need to show that $x \in \overline{S} = S \cup S'$. If $x \in S$, then clearly $x \in \overline{S}$ so assume that $x \notin S$, and thus $x \in \partial S$. But then if U is an open set containing x, then U contains a point from S (and from $X - S$, though this is irrelevant here), but then $U \cap (S - \{x\}) \neq \emptyset$ since $x \notin S$. Thus x is a cluster point, as needed. □

Corollary 3.1 *Let X be a topological space, and $S \subseteq X$ a subset. Then*

1. The interior $\text{int}(S)$ is the largest open set in X that is contained in S. In particular, $\text{int}(S)$ is open.
2. S is open in X if, and only if, $S = \text{int}(S)$. In other words, S is open precisely when all its points are interior.
3. The exterior $\text{ext}(S)$ is the largest open set in X that is disjoint from S. In particular, $\text{ext}(S)$ is open.
4. S is closed in X if, and only if, $X - S = \text{ext}(S)$.

5. The closure $\overline{S}$ is the smallest closed set in X that contains S. In particular, $\overline{S}$ is closed.
6. S is closed in X if, and only if, $S' \subseteq S$ (equivalently, if $\overline{S} = S$).

3.2.2 Sequences and Convergence

We now turn to look at sequences in topological spaces, what it means for such sequences to converge, and how this notion relates to the topological concepts given above.

Definition 3.14 Let $\{x_m\}_{m\geq 1}$ be a sequence of points in a topological space X. A point $x_0 \in X$ is said to be a *limit* of the sequence $\{x_m\}_{m\geq 1}$ if for every open set $U \subseteq X$ with $x_0 \in U$, there exists an $N \in \mathbb{N}$ such that

$$m > N \implies x_m \in U.$$

The sequence is then said to *converge* to x_0 and to be a *convergent sequence.*

Just as continuity of functions can be detected using a basis, so can cluster points and limit points be so detected.

Proposition 3.10 *In a topological space X with a local basis $\mathscr{B}$ at a point x_0:*

1. *x_0 is a cluster point of a subset $S \subseteq X$ if, and only if, $B \cap (S - \{x_0\}) \neq \emptyset$, for all basis elements $B \in \mathscr{B}$.*
2. *A sequence $\{x_m\}_{m\geq 1}$ converges to x_0 if, and only if, for every $B \in \mathscr{B}$ there exists an $N \in \mathbb{N}$ such that*
$$m > N \implies x_m \in B.$$

Proof We only prove the first assertion, as the two proofs are very similar. If x_0 is a cluster point and $B \in \mathscr{B}$ is a basis element, then, in particular, B is open and $x_0 \in B$, which implies that $B \cap (S - \{x_0\}) \neq \emptyset$. Conversely, suppose the stated condition holds and let U be an arbitrary open set with $x_0 \in U$. By definition of basis at a point, there exists a basis element $B \in \mathscr{B}$ with $x_0 \in B \subseteq U$. By hypothesis then $B \cap (S - \{x_0\}) \neq \emptyset$ and since $U \cap (S - \{x_0\}) \supseteq B \cap (S - \{x_0\})$, the proof is complete. □

Example 3.10 Consider the Euclidean topology on $\mathbb{R}$ and recall that a local basis at x_0 is given by $\{B_\varepsilon(x_0)\}_{\varepsilon>0}$. Thus, a sequence $\{x_m\}_{m\geq 1}$ in $\mathbb{R}$ converges to $x_0 \in \mathbb{R}$ if, and only if, for every $\varepsilon > 0$ there exists an $N \in \mathbb{N}$ such that

$$m > N \implies x_m \in B_\varepsilon(x_0),$$

or, in other words: if $m > N$, then $|x_m - x| < \varepsilon$. We thus see that convergence in the topological sense given above is equivalent to convergence of sequences in $\mathbb{R}$ in the usual sense.

It is very reassuring that the notion of convergence can be captured topologically, but one should be cautious not to expect the familiar behaviour of limits in $\mathbb{R}$ to remain valid in arbitrary topological spaces, as the following example illustrates.

Example 3.11 Let X be an arbitrary set endowed with the indiscrete topology. Since the only open sets are $\emptyset$ and X, it follows immediately that any sequence $\{x_m\}_{m\geq 1}$ in X converges and that *any* element $x \in X$ is a limit. Consequently, a sequence may converge to more than one point; uniqueness of limits is lost.

Given a subset S of a topological space X, we can entertain (at least) two reasonable formalizations for the informal idea of the set of all points that are infinitesimally close to S. One possibility is given by the closure $\overline{S} = S \cup S'$, where we adjoin to S all of its derived points. A second possibility is obtained by considering the set of all limit points of sequences in S. A-priorly it is not clear how these two possibilities compare, and we now turn to investigate this question. We limit the discussion just to the point of developing enough theory to suit the needs of this book. It should be noted that this is just the tip of the iceberg. The important observation is that, generally, sequences do not suffice to capture cluster points (but either nets or filters, two concepts we will not discuss, do). Having said that, we do identify below a broad enough class of topological spaces in which the limit behaviour is more intuitive, namely spaces satisfying the first countability axiom.

Theorem 3.6 *Let X be a topological space and $S \subseteq X$ a subset. If a non-eventually-constant sequence $\{s_m\}_{m\geq 1}$ in S converges to $x_0 \in X$, then x_0 is a cluster point of the subset S.*

Proof Suppose that U is an open set and $x_0 \in U$. Since $\{s_m\}_{m\geq 1}$ converges to x_0, there exists an $N \in \mathbb{N}$ such that $s_m \in U$ for all $m > N$, and since the sequence is not eventually constant, an $m > N$ exists with $s_m \neq x_0$. In particular, $U \cap (S - \{x_0\}) \neq \emptyset$, and thus x_0 is a cluster point of S. □

Theorem 3.7 *If $f : X \to Y$ is a continuous function between topological spaces and $x_m \to x_0$ in X, then $f(x_m) \to f(x_0)$ in Y.*

Proof Given an open set $U \subseteq Y$ with $f(x_0) \in U$, we have that $x_0 \in f^{-1}(U)$ and thus there exists an $N \in \mathbb{N}$ such that

$$m > N \implies x_m \in f^{-1}(U)$$

and thus

$$m > N \implies f(x_m) \in U$$

as required. □

Remark 3.6 The converses of these two theorems are generally false. To see that, it suffices to note that if X is an uncountable set endowed with the cocountable topology, then the only convergent sequences are the eventually constant ones, while every point $x \in X$ is a cluster point of any cofinite set S (i.e. one where where $X - S$ is finite).

3.2.3 Second Countable and First Countable Spaces

We present now two axioms out of a class of properties called countability axioms. The conditions we present are quite strong, and have pleasant consequences as we will see (particularly to the behaviour of limits and cluster points). While many topological spaces fail to satisfy either of the countability axioms we present, the class of those that do is broad enough to supply one with plenty of spaces that topologically are closer to one's intuition than their wilder topological cousins are.

Definition 3.15 A topological space X satisfies the *second axiom of countability* or is said to be *second countable* if there is a countable basis for the topology on X.

Thus, second countability means that there exist a countable index set I and a family $\{U_i\}_{i\in I}$ of open sets in X such that any set U can be expressed as

$$U = \bigcup_{j\in J_U} U_j$$

for some $J_U \subseteq I$.

Example 3.12 The space $\mathbb{R}$ with the Euclidean topology is second countable. To see that, we exhibit a countable basis for the topology, namely the collection $\{B_{1/n}(q)\}_{q\in\mathbb{Q},n\geq 1}$. First, note that this is a countable collection, as it is indexed by the set $\mathbb{Q}\times(\mathbb{N}-\{0\})$ (see, e.g., Sect. 1.2.13 of the Preliminaries). Next, given an open set $U\subseteq\mathbb{R}$ and a point $x\in U$, there exists a $\delta>0$ such that $B_\delta(x)\subseteq U$, and let $n\in\mathbb{N}$ satisfy $1/n<\delta/2$. Since the rationals are dense in the reals, we may find a rational number q for which $|x-q|<1/n$. It now easily follows that $x\in B_{1/n}(q)\subseteq B_\delta(x)\subseteq U$, as required by the definition of basis.

Definition 3.16 A subset $S\subseteq X$ in a topological space is said to be *dense* in X if $\overline{S}=X$. The space X is called *separable* if it contains a countable dense subset.

A familiar example of a separable topological space is $\mathbb{R}$ with the Euclidean topology. Indeed, it is easily verified that the countable set $\mathbb{Q}$ is dense in $\mathbb{R}$. In fact, $\mathbb{R}$ is also seen to be separable by the following general result.

Theorem 3.8 *Every second countable space X is separable.*

Proof Let $\mathscr{B}$ be a countable basis for the topology. For each $B\in\mathscr{B}$ choose an arbitrary point $x_B\in B$ (if $B=\emptyset$, simply skip it). The set $S=\{x_B\mid B\in\mathscr{B}\}$ is clearly countable so it remains to show that it is dense. Indeed, fix a point $x\in X$. To show that x is in the closure of S suppose $x\notin S$ and let U be an open set with $x\in U$. We now need to show that $U\cap(S-\{x\})\neq\emptyset$. Indeed, by definition, there exists a basis element B with $x\in B\subseteq U$. But then x_b must be found in $U\cap(S-\{x\})$, establishing the claim. □

Many spaces of interest are not second countable but they do satisfy a (considerably) weaker condition known as the first axiom of countability. Such spaces, as we will shortly see, exhibit a general behaviour that in some respect is quite close to what our intuition about convergence and closures in $\mathbb{R}$ dictates.

Definition 3.17 A topological space X satisfies the *first axiom of countability* or is said to be *first countable* if there exists, at every point $x \in X$, a countable local basis.

Evidently a second countable space is also first countable, but not, generally, vice versa.

Recall that the converses of theorems 3.6, and 3.7 are not generally true in arbitrary topological spaces. For first countable spaces however, we have the following pleasing result.

Theorem 3.9 *Let X be a first countable space, that is, at each point $x_0 \in X$ there exists a countable local basis $\mathscr{B}_{x_0} = \{B^x_m\}_{m \geq 1}$ at x_0.*

1. *For a subset $S \subseteq X$, a point $x_0 \in X$ is in $\overline{S}$ if, and only if, $s_m \to x_0$ for a sequence $\{s_m\}_{m \geq 1}$ in S.*
2. *A subset $F \subseteq X$ is closed if, and only if, F contains all limit points of sequences in it.*
3. *A function $f : X \to Y$ to any topological space Y is continuous if, and only if, it preserves limits of sequences, i.e.,*

$$x_m \to x_0 \implies f(x_m) \to f(x_0)$$

for all sequences $\{x_m\}_{m \geq 1}$ in X.

Proof

1. By Theorem 3.6, every limit point of a non-eventually constant sequence in S is a cluster point of S, and thus is in $\overline{S}$. The limits of eventually constant sequences are, of course, in S and thus also in $\overline{S}$. Suppose now that $x_0 \in X$ is in $\overline{S}$ and, of course, the interesting case is when $x_0 \notin S$. In that case x_0 is a cluster point of S. Therefore, for every $m \geq 1$ the open set $B^{x_0}_1 \cap \cdots \cap B^{x_0}_m$ intersects $S - \{x_0\}$ non-trivially, and let s_m be an arbitrary element in the intersection. Clearly the sequence $\{s_m\}_{m \geq 1}$ is in S and its limit is x_0. Indeed, given any open set $U \subseteq X$ with $x_0 \in U$, there exist a basis element $B^{x_0}_N$ with $x_0 \in B^{x_0}_N \subseteq U$. But then

$$m > N \implies s_m \in B^{x_0}_m \subseteq B^{x_0}_N \subseteq U.$$

2. By Corollary 3.1, F is closed if, and only if, $\overline{F} = F$, and we just proved that the elements of $\overline{F}$ coincide with limit points of sequences in F.
3. By Theorem 3.7, if $f : X \to Y$ is continuous, then it preserves limits of sequences. Suppose thus that $f : X \to Y$ preserves limits of sequences, and our aim is to show that f is continuous. We need to show that $f^{-1}(U)$ is open whenever $U \subseteq Y$ is open, or (equivalently!) that $f^{-1}(F)$ is closed whenever $F \subseteq Y$ is closed. It suffices to show that $f^{-1}(F)$ contains its limit points, so suppose that

$$x_m \to x_0$$

in X with $\{x_m\}_{m \geq 1}$ a sequence in $f^{-1}(F)$, and we will show that $x_0 \in f^{-1}(F)$. Indeed, f preserves limits of sequences and thus $f(x_m) \to f(x_0)$, but then $f(x_0)$

is a limit point of a sequence in F, and since the latter is closed, Theorem 3.6 implies that $f(x_0) \in F$, showing that $x_0 \in f^{-1}(F)$. □

Exercises

Exercise 3.11 Prove that the boundary $\partial(S)$ of any subset S of a topological space X is a closed set.

Exercise 3.12 Prove that $\partial S = \partial(X - S)$ for any subset S of a topological space X.

Exercise 3.13 Prove that a subset S of a topological space X is closed if, and only if, it contains its boundary.

Exercise 3.14 Prove that a set S of a topological space is open if, and only if, it is disjoint from its boundary.

Exercise 3.15 Prove that $\overline{\overline{S}} = \overline{S}$ for all subsets S of a topological space X.

Exercise 3.16 Find an example of a topological space X and a non-empty subset $S \subseteq X$ such that

$$\text{int}(S) = \text{ext}(S) = \emptyset.$$

Exercise 3.17 Let X be an uncountable set endowed with the cocountable topology. Show that the only convergent sequences in X are the eventually constant ones, while for every cofinite set $S \subseteq X$ (i.e., $X - S$ is finite) $S' = X$.

Exercise 3.18 Construct a first countable space that is not second countable.

Exercise 3.19 Prove that a countable first countable space is second countable. That is, if X is a countable set endowed with a first countable topology, then it is in fact a second countable topology.

Exercise 3.20 Prove that if X is a second countable topological space, then $|\tau_X|$, the cardinality of τ_X, is at most $|\mathbb{R}|$, the cardinality of the real numbers. Does the converse hold? Does the same result hold if second countability is replaced by first countability?

3.3 Constructing Topologies

In this section we discuss two ways of constructing topologies. The first situation is where one has an arbitrary set and a collection of subsets of it that one would like to have as the open sets in a topology on X. The second scenario concerns a collection of functions one would like to force to be continuous. This construction is used to obtain products, coproducts, and quotients for topological spaces.

3.3.1 Generating Topologies

We have seen above that it is at times convenient to have a manageable basis for a topology since one can effectively work with the basis elements rather than the entire topology. Given an arbitrary set X, without any prior notion of topology, and a collection $\mathscr{B}$ of subsets of X, it is natural to ask whether there exists a topology τ on X for which $\mathscr{B}$ is a basis, and, if such a topology exists, is it unique. The answers are quite decisive.

Theorem 3.10 (Basis Generating Topology) *A collection $\mathscr{B}$ of subsets of a set X is a basis for a topology τ on X if, and only if, the following conditions are met.*

1. *For every $x \in X$ there exists a $B \in \mathscr{B}$ such that $x \in B$. Equivalently,*

$$X = \bigcup_{B \in \mathscr{B}} B.$$

2. *For all $B_1, B_2 \in \mathscr{B}$, if $x \in B_1 \cap B_2$, then there exists $B_3 \in \mathscr{B}$ such that*

$$x \in B_3 \subseteq B_1 \cap B_2.$$

Moreover, when the conditions are met, the topology τ is unique, namely τ is the smallest topology on X such that every $B \in \mathscr{B}$ is open.

Proof Let us call a subset $U \subseteq X$ open if for every $x \in U$ there exists a $B \in \mathscr{B}$ such that

$$x \in B \subseteq U.$$

Equivalently, $U \subseteq X$ is open if U is a union of elements from B. We show that the collection τ of all such open sets forms a topology on X. The empty set $\emptyset$ is vacuously open since it has no points, and the set X itself is open by condition 1. Stability under arbitrary unions is straightforward, so let us establish stability under finite intersections. Let $U_1, U_2 \subseteq X$ be open and we need to show that $U = U_1 \cap U_2$ is open. Indeed, if $x \in U$, then $x \in U_1$ and thus there is a $B_1 \in \mathscr{B}$ such that $x \in B_1 \subseteq U_1$. Similarly, there is a $B_2 \in \mathscr{B}$ with $x \in B_2 \subseteq U_2$. As $x \in B_1 \cap B_2$, condition 2 gives us an element $B_3 \in \mathscr{B}$ with

$$x \in B_3 \subseteq B_1 \cap B_2 \subseteq U_1 \cap U_2 = U,$$

as needed.

The claim that $\mathscr{B}$ is a basis for τ is immediate from the construction. As for the minimality claim about τ, note that if τ' is any topology on X for which all $B \in \mathscr{B}$ are open, then any $U \in \tau$, being a union of elements from B, must also be open in τ'. In other words, $\tau \subseteq \tau'$. The uniqueness of τ now follows too. □

We may now pose the exact same question as above, replacing basis by subbasis. Thus, given an arbitrary collection $\mathscr{B}$ of subsets of X, is there a topology τ on X for which $\mathscr{B}$ is a subbasis, and is this topology unique.

Theorem 3.11 (Subbasis Generating Topology) *A collection $\mathscr{B}$ of subsets of a set X is a subbasis for a topology τ on X if, and only if,*

$$\bigcup_{B \in \mathscr{B}} B = X.$$

Moreover, in that case the topology τ is unique, namely τ is the smallest topology on X such that every $B \in \mathscr{B}$ is open.

Proof One may reduce the proof to that of the previous result by noting that the collection

$$\mathscr{B}' = \{B_1 \cap \cdots \cap B_m \mid B_1, \ldots, B_m \in \mathscr{B},\ m \geq 0\},$$

consisting of all finite intersections of elements of $\mathscr{B}$, does satisfy the conditions of being a basis for a topology, and thus the previous result may be applied to $\mathscr{B}'$, giving rise to a topology τ. In other words, we declare a subset $U \subseteq X$ to be open if it is an arbitrary union of finite intersections of elements from $\mathscr{B}$. The details are left to the reader. □

The results above, and their proofs, provide one with powerful tools for creating topologies. Any topology constructed from a collection $\mathscr{B}$ by the above techniques is said to be *generated* by $\mathscr{B}$.

The following quite general result is used often to produce interesting topologies. The result states, roughly, that given any collection of functions with a common domain or codomain, a canonical topology exists rendering all the given functions continuous. We remark that, in a sense, topological spaces exists as the servants of continuous functions; we define topological spaces since we are interested in continuous mappings. This result thus allows one to tailor a suitable topology from a given collection of functions.

Proposition 3.11 (Topologies Induced by Functions)

1. *Let X be a set and $\{f_i : X \to X_i\}_{i \in I}$ a non-empty collection of functions with each X_i a topological space. There exists then a unique smallest topology on X such that every $f_i : X \to X_i$ is continuous.*
2. *Let X be a set and $\{f_i : X_i \to X\}_{i \in I}$ a non-empty collection of functions with each X_i a topological space. There exists then a unique largest topology on X such that every $f_i : X_i \to X$ is continuous.*

Proof

1. Suppose that τ is a topology on X such that each $f_i : X \to X_i$ is continuous. That means that $f_i^{-1}(U) \in \tau$ for each $U \in \tau_{X_i}$. In other words $\tau \supseteq \mathscr{B}$ where

$$\mathscr{B} = \{f_i^{-1}(U) \mid i \in I, U \in \tau_{X_i}\}$$

and so we see that we must define τ to be the topology generated by $\mathscr{B}$. This is feasible since $\mathscr{B}$ is easily seen to satisfy the condition of Theorem 3.11.

2. Suppose that τ is a topology on X such that each $f_i : X_i \to X$ is continuous. That means that τ can not contain any subset $S \subseteq X$ for which there exists an $i \in I$ with $f_i^{-1}(S) \notin \tau_{X_i}$. So, to obtain the largest possible topology on X, we consider the collection

$$\tau = \{S \subseteq X \mid f_i^{-1}(S) \in \tau_{X_i} \text{ for all } i \in I\}.$$

If we can show that τ is indeed a topology, then the proof is complete. Notice that

$$\tau = \bigcap_{i \in I} \{S \subseteq X \mid f_i^{-1}(S) \in \tau_{X_i}\}$$

and that (the reader is invited to prove) the intersection of a family of topologies on X is again a topology on X. Thus we only need to establish that the collection $\{S \subseteq X \mid f_i^{-1}(S) \in \tau_{X_i}\}$, for a single $i \in I$, is a topology. This follows immediately from well-known properties of the inverse image function, as the reader may verify. □

When a topology τ is constructed as above, we say that it is generated by the given collection of functions. We note at once that it is quite possible that the collection of functions consists of just one single function.

Remark 3.7 Definition 3.11 of the subspace topology on a subset Y of a topological space X is already an example of this method of constructing topologies. Indeed, the subspace topology is precisely the topology generated by the single function $i : Y \to X$, the inclusion function (see Proposition 3.8).

3.3.2 Coproducts, Products, and Quotients

The rest of this section utilizes the constructions above in three particular scenarios.

Definition 3.18 Let X and Y be topological spaces and assume $X \cap Y = \emptyset$. The topology on $X \cup Y$ generated by $\{i_X : X \to X \cup Y, i_Y : Y \to X \cup Y\}$, where i_X and i_Y are the inclusion functions, is called the *coproduct topology* on $X \cup Y$, and the space $X \cup Y$ is called the *coproduct* of X and Y.

Examples are abundant, but some care is required in order to hone the intuition. For instance, in $\mathbb{R}$ with the Euclidean topology, the two subspaces $\mathbb{Q}$ and $\mathbb{R} - \mathbb{Q}$ are clearly disjoint, so one can form their coproduct which would give a topology on $\mathbb{R}$ which is not the Euclidean topology.

The situation easily generalizes to infinitely many spaces, as follows. Suppose $\{X_i\}_{i\in I}$ is a non-empty collection of topological spaces that are pairwise disjoint. The *coproduct* of all the X_i is then the set

$$X = \bigcup_{i\in I} X_i$$

endowed with the topology generated by the family of inclusions $\{X_i \to X\}_{i\in I}$. As an example, a topological space X is discrete if, and only if, it is the coproduct of its singleton subspaces.

Definition 3.19 Let X and Y be two topological spaces, and consider the cartesian product $X \times Y$. The topology generated by the set of projections

$$\{\pi_X : X \times Y \to X, \pi_Y : X \times Y \to Y\}$$

is called the *product topology* on $X \times Y$, and the space $X \times Y$ is called the *cartesian product* or simply the *product* of X and Y.

This situation also naturally extends to the infinite case. If $\{X_i\}_{i\in I}$ is a non-empty family of topological spaces, then their cartesian product is the set X, the set-theoretic cartesian product of the X_i, endowed with the topology generated by the family of projections $\{\pi_i : X \to X_i\}_{i\in I}$.

Let us consider the case of a product of countably many spaces $\{X_m\}_{m\geq 1}$. The set-theoretic cartesian product $X = X_1 \times \cdots \times X_m \times \cdots$ consists of all sequences $x = (x_1, \dots, x_m, \dots)$ with $x_m \in X_m$ for each $m \geq 1$. The projection $\pi_m : X \to X_m$ is given by $\pi_m(x) = x_m$. Following the definitions above, it follows that the product topology on X is the one generated by the sets of the form $U_1 \times \cdots \times U_m \times \cdots$ where each U_m is an open set in X_m and $U_m = X_m$ for all but finitely many m.

Example 3.13 A familiar example is $\mathbb{R}^n$, the n-fold product of the topological space $\mathbb{R}$ with itself, when given the Euclidean topology. For $n \geq 2$ the resulting product topology on $\mathbb{R}^n$ is also called the Euclidean topology. Another example is $\mathbb{R}^\infty$ as a countable product of $\mathbb{R}$ with itself.

The final construction we present is that of the quotient topology. Let X be a set with an equivalence relation on it. Recall (e.g., Sect. 1.2.15 of the Preliminaries) that there is then the associated quotient set $X/{\sim} = \{[x] \mid x \in X\}$ of all equivalence classes $[x]$, and the canonical projection $\pi : X \to X/{\sim}$ given by $\pi(x) = [x]$. When this situation is enriched by the presence of a topology on X, it is natural to seek out a topology on $X/{\sim}$ as well. With the tools we now have, this is immediate.

Definition 3.20 Let X be a topological space and $\sim$ an equivalence relation on the set X. The topology on $X/{\sim}$ generated by the projection function $\pi : X \to X/{\sim}$ is called the *quotient topology* on $X/{\sim}$, and the space $X/{\sim}$ is called the *quotient of X modulo* $\sim$.

Example 3.14 Consider $[0, 1]$ as a subspace of $\mathbb{R}$ with its standard topology. Define the equivalence relation $\sim$ on $[0, 1]$ by $x \sim y$ precisely when $x = y$ or $x, y \in \{0, 1\}$. The quotient space $[0, 1]/\sim$ is homeomorphic to the space $S^1 = \{x \in \mathbb{R}^2 \mid \|x\| = 1\}$ considered as a subspace of $\mathbb{R}^2$ with the Euclidean topology. Informally, $\sim$ glues the two ends of the interval $[0, 1]$, resulting in a circle.

Exercises

Exercise 3.21 Let X be a set and $\{\tau_i\}_{i \in I}$ a collection of topologies on X. Prove that

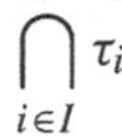

$$\bigcap_{i \in I} \tau_i$$

is a topology as well. On the contrary, prove that the union of two topologies on X need not be a topology.

Exercise 3.22 Let X be a set and $\mathscr{B}$ a collection of subsets of X satisfying the conditions of Theorem 3.11. Prove that the intersection of all topologies τ with $\mathscr{B} \subseteq \tau$ is precisely the topology generated by $\mathscr{B}$.

Exercise 3.23 Prove that the collection $\{[a, b) \mid a, b \in \mathbb{R},\ a < b\}$ is a basis for a topology on $\mathbb{R}$, known as the lower-limit topology. When $\mathbb{R}$ is equipped with this topology it is called the *Sorgenfrey line*, denoted by $\mathbb{R}_l$. We will denote $\mathbb{R}$ with the Euclidean topology by $\mathbb{R}_E$. Compare these two topologies on $\mathbb{R}$ and investigate the meaning of continuity of functions $f : \mathbb{R}_l \to \mathbb{R}_E$.

Exercise 3.24 Let X be a topological space. Prove that X is discrete if, and only if, X is the coproduct of all its singleton subspaces.

Exercise 3.25 Prove that for all topological space X, Y, Z, the spaces $X \times (Y \times Z)$, $X \times Y \times Z$, and $(X \times Y) \times Z$ are homeomorphic. Is there any situation in which these spaces are equal?

Exercise 3.26 Prove or find a counterexample to the claim that for all topological spaces X and Y, any subspace S of $X \times Y$ is necessarily of the form of the topological product $S_X \times S_Y$ for some subspaces $S_X \subseteq X$ and $S_Y \subseteq Y$.

Exercise 3.27 Let X and Y be topological spaces and consider the relation on the set $X \times Y$ given as follows. Declare $(x, y) \sim (x', y')$ precisely when $x = x'$ (obviously an equivalence relation). Prove that when $X \times Y$ is endowed with the product topology, the quotient space $(X \times Y)/\sim$ is homeomorphic to X. Can equality ever hold?

Exercise 3.28 Consider $\mathbb{R}$ with the Euclidean topology. Prove that the collection $\{U_1 \times \cdots \times U_m \times \cdots \mid U_1, \ldots, U_m \subseteq \mathbb{R} \text{ are open}\}$ is a basis for a topology on $\mathbb{R}^\infty$. This topology is called the *box* topology. Of the box topology and the product topology on $\mathbb{R}^\infty$, which one is finer?

Exercise 3.29 Let X be a topological space and $\sim$ an equivalence relation on X. Prove that a set $U \subseteq X/\sim$ (which is a collection of equivalence classes, thus of subsets of X) is open in the quotient topology if, and only if,

$$\bigcup_{[x]\in U} [x]$$

is open in X.

Exercise 3.30 Construct a quotient space that is homeomorphic to a torus in $\mathbb{R}^3$.

3.4 Separation and Connectedness

Separation properties of a topological space relate to the ability of the open sets to separate distinct points, a point from a set, or two disjoint sets. We only consider one such property, the Hausdorff separation property, which is the strongest of the separation axioms pertaining to the ability to separate distinct points. We then turn our attention to the notion of connectivity for a topological space, a notion that is somewhat more subtle than one might initially expect.

3.4.1 The Hausdorff Separation Property

If X is endowed with the indiscrete topology, then the open sets (of which there are only two) are, in a sense, blind to the individual points in the space. The open sets are simply not sufficiently refined to be able to distinguish between different points. This is an extreme situation of course and it goes against one's intuition of how distinct points should (in some sense) behave. The following separation property immediately brings some of the familiarity back.

Definition 3.21 A topological space X is said to satisfy the *Hausdorff separation property*, or simply to be *Hausdorff*, if for all distinct points $x, y \in X$ there exist disjoint open sets U and V such that $x \in U$ and $y \in V$. Such open sets are said to separate x and y.

Remark 3.8 The Hausdorff separation property is also commonly referred to as T_2, indicating its position in a hierarchy of several separation axioms, starting with T_0, the weakest one. We will not delve into this issue here.

Example 3.15 $\mathbb{R}$ with the Euclidean topology is Hausdorff. Indeed, if $x \neq y$ are real numbers, then setting

$$\delta = \frac{|x-y|}{2} > 0$$

one easily sees that

$$B_\delta(x) \cap B_\delta(y) = \emptyset$$

thus clearly exhibiting open sets that separate x and y.

Example 3.16 An example of a non-Hausdorff space is given by the Sierpinski space $\mathbb{S} = \{0, 1\}$. Its two points clearly can not be separated, since the topology is given by $\{\emptyset, \{1\}, \{0, 1\}\}$. In fact, a finite topological space X is Hausdorff if, and only if, it is discrete. Indeed any discrete space is clearly Hausdorff, since any two distinct points x and y are separated by the open sets $\{x\}$ and $\{y\}$. In the other direction, suppose X is finite and Hausdorff, and choose some $x \in X$. For each $y \in X - \{x\}$ we may find disjoint open sets U_y and V_y such that $x \in U_y$ and $y \in V_y$. Then, the intersection

$$U = \bigcap_{y \in X - \{x\}} U_y$$

is an open set (this is where the finiteness assumption on X is used), and it contains x. Further, since $U \subseteq U_y$, it follows that

$$U \cap V_y \subseteq U_y \cap V_y = \emptyset$$

and thus U is an open set that contains x but does not contain any $y \in X - \{x\}$. In other words, $U = \{x\}$, and thus we proved that $\{x\}$ is an open set. Since x was arbitrary, it follows that X is discrete.

One immediate consequence of the Hausdorff property is the following result.

Theorem 3.12 *A convergent sequence $\{x_m\}_{m \geq 1}$ in a Hausdorff space X converges to a unique limit.*

Proof Suppose that $\{x_m\}_{m \geq 1}$ converges to two different points x and y. There exist, by the Hausdorff property, two disjoint open sets U_x and U_y with $x \in U_x$ and $y \in Y_y$. By Definition 3.14, there exist $N_x, N_y \in \mathbb{N}$ such that

$$m > N_x \implies x_m \in U_x$$

$$m > N_y \implies x_m \in U_y.$$

But then, for $m > \max\{N_x, N_y\}$, we have that $x_m \in U_x \cap U_y$, a contradiction since U_x and U_y are disjoint. □

3.4.2 Path-Connected and Connected Spaces

The second property we discuss is that of connectedness which, to immediately dispel any misconception, is *not* opposing the Hausdorff separation property (or any other separation property). There are two notions to discuss—path connectivity and connectivity—the former more immediately intuitive than the latter, and thus is the one we present first.

Definition 3.22 A space X is said to be *path-connected* if for all points $x, y \in X$ there exists a continuous function $\gamma : [0, 1] \to X$ with $\gamma(0) = x$ and $\gamma(1) = y$.

Here the topology given to the interval $[0, 1]$ is the subspace topology induced by the Euclidean topology on $\mathbb{R}$. A continuous function $\gamma : [0, 1] \to X$ is naturally thought of as a *path* in X with initial point $\gamma(0)$ and final destination $\gamma(1)$. Thus, X is path-connected if any two points in the space can be connected by a path in the space.

Example 3.17 The space $\mathbb{R}^n$ is path-connected for all $n \geq 1$. Indeed, given points $x = (x_1, \ldots, x_n)$ and $y = (y_1, \ldots, y_n)$, it is easily verified that $\gamma : [0, 1] \to \mathbb{R}^n$ given by

$$\gamma(t) = t(y_1, \ldots, y_n) + (1 - t)(x_1, \ldots, x_n)$$

is a path connecting x and y. The same argument shows that any subset of $\mathbb{R}^n$ of the form

$$[a_1, b_1] \times \cdots \times [a_n, b_n],$$

a product of intervals, is path-connected. In particular, any interval $[a, b]$ in $\mathbb{R}$ is path-connected. More generally, any convex subset of $\mathbb{R}^n$ is path-connected. The reader is invited to verify that another example of a connected space is $S = \{x \in \mathbb{R}^n \mid \|x\| = 1\}$, the unit sphere, viewed as a subspace of $\mathbb{R}^n$ with the Euclidean topology, provided that $n > 1$.

An example of a space that is not path-connected is $\mathbb{Q}$ as a subspace of $\mathbb{R}$ with the Euclidean topology. In fact, no two distinct points $x, y \in \mathbb{Q}$ can be connected by a path. To see that, one can show that any path $\gamma : [0, 1] \to \mathbb{Q}$ is constant. However, we establish the same result by utilizing the concept of connectivity which we now turn to.

Definition 3.23 A topological space X is said to be *disconnected* if there exists non-empty open sets U and V such that $U \cap V = \emptyset$ and $U \cup V = X$. A space X is said to be *connected* if it is not disconnected.

Theorem 3.13 *Any closed interval $[a, b]$, as a subspace of $\mathbb{R}$ with the Euclidean topology, is connected.*

Proof Without loss of generality let us assume that $[a, b] = [0, 1]$. Suppose that $[0, 1]$ is not connected. Then there exist non-empty open subsets U and V in the subspace topology such that $U \cup V = [a, b]$ and $U \cap V = \emptyset$. Since a singleton set is not open in the subspace topology on $[0, 1]$ it follows that not U nor V is a singleton set. In particular, we may assume that there exist $a \in U$ and $b \in V$ with $0 < a < b < 1$. The set $S = \{x \in U \mid x < b\}$ is non-empty and bounded above, and so admits a supremum s, and notice that $0 < s < 1$.

Now, since $[0, 1] = U \cup V$ we must have that either $s \in U$ or $s \in V$. If $s \in U$ then $s < b$ and, U being open, there exists an $a' \in U$ with $s < a' < b$, contrary to the choice of s. On the other hand, if $s \in V$ then, V being open, there exist an $\varepsilon > 0$ with $(s - \varepsilon, s + \varepsilon) \subseteq V$, again contrary to the choice of s. The assumption of the existence of U and V as above thus leads to a contradiction, and so $[0, 1]$ is connected. □

Remark 3.9 The intuition behind the concept of connectedness is perhaps better explained by referring to the word 'samenhangend', the term used for this property in the Dutch language. The literal meaning of 'samenhangend' is 'hanging together', and it is this topological quality that the definition of connectedness captures. For instance, it is intuitively clear that if a space is path-connected, then it must be 'hanging together'. This is indeed the case.

Lemma 3.1 *If a topological space X is path-connected, then it is connected.*

Proof Suppose that X is disconnected. That means that there exist disjoint non-empty open sets $U, V \subseteq X$ such that

$$U \cup V = X.$$

Choose points $x \in U$ and $y \in V$ and, utilizing the assumption that the space X is path-connected, let $\gamma : [0, 1] \to X$ be a path from x to y. Since γ is continuous, it follows that $\gamma^{-1}(U)$ and $\gamma^{-1}(V)$ are open sets in $[0, 1]$. Further,

$$\gamma^{-1}(U) \cap \gamma^{-1}(V) = \gamma^{-1}(U \cap V) = \gamma^{-1}(\emptyset) = \emptyset$$

and

$$\gamma^{-1}(U) \cup \gamma^{-1}(V) = \gamma^{-1}(U \cup V) = \gamma^{-1}(X) = [0, 1].$$

Lastly, since $x = \gamma(0)$ and $x \in U$ it follows that $0 \in \gamma^{-1}(U)$, and in particular $\gamma^{-1}(U) \neq \emptyset$. Similarly one shows that $\gamma^{-1}(V) \neq \emptyset$.

The conclusion is thus that $\gamma^{-1}(U)$ and $\gamma^{-1}(V)$ separate $[0, 1]$, a contradiction with the fact that $[0, 1]$ was shown to be a connected subspace of $\mathbb{R}$. We conclude that X is connected. □

Example 3.18 We now return to the example of $\mathbb{Q}$ and show that it is disconnected, and thus also not path-connected. Indeed, let α be an arbitrary irrational number and let $U = \{x \in \mathbb{Q} \mid x < \alpha\}$ and $V = \{x \in \mathbb{Q} \mid x > \alpha\}$. Clearly, $U \cap V = \emptyset$ and $U \cup V = \mathbb{Q}$, and thus all that is left to show is that U and V are open. Indeed, given any $x \in U$, setting $\delta = (\alpha - x)/2$, one easily sees that $(x - \delta, x + \delta) \cap \mathbb{Q} \subseteq U$. But this shows that U is open in the subspace topology on $\mathbb{Q}$, as required. The claim that V, too, is open follows similarly.

Theorem 3.14 *If $f : X \to Y$ is a surjective continuous mapping between topological spaces and X is connected, then so is Y.*

Proof Suppose that Y is disconnected, namely that $Y = U \cup V$ for some non-empty open sets $U, V \subseteq Y$ which further satisfy $U \cap V = \emptyset$. Note that

$$X = f^{-1}(Y) = f^{-1}(U) \cup f^{-1}(V)$$

and that both $f^{-1}(U)$ and $f^{-1}(V)$ are open in X. Further,

$$f^{-1}(U) \cap f^{-1}(V) = f^{-1}(U \cap V) = f^{-1}(\emptyset) = \emptyset.$$

Since X is connected, we may conclude that either $f^{-1}(U)$ or $f^{-1}(V)$ is empty. But, since f is surjective, that is impossible and we conclude that Y is connected. □

Exercises

Exercise 3.31 Find a topological space X which is not Hausdorff, and a quotient of X which is Hausdorff. Also, find a topological space X which is Hausdorff, and a quotient of X which is not Hausdorff.

Exercise 3.32 Call two points x, y in a topological space X *inseparable* if for all open sets U, $x \in U \iff y \in U$. Prove that inseparability is an equivalence relation on X. Is the quotient space $X/\!\sim$ Hausdorff?

Exercise 3.33 Given topological spaces X and Y, prove that $X \times Y$ is Hausdorff if, and only if, each of X and Y is Hausdorff.

Exercise 3.34 Prove that a topological space X is Hausdorff if, and only if, the diagonal set $\{(x, x) \mid x \in X\}$ is closed in the product space $X \times X$.

Exercise 3.35 Prove that each of the Hausdorff separation property, connectivity, and path-connectivity is a topological invariant. Use a connectivity argument to show that $\mathbb{R}$ and $\mathbb{R}^2$, each with the Euclidean topology, are not homeomorphic. (Hint: improve some of these topological invariants.)

Exercise 3.36 Let X be a topological space, and $S \subseteq X$ a subset. Prove that if S (as a subspace) is connected, then so is its closure $\bar{S}$.

Exercise 3.37 Characterize all of the connected subsets of $\mathbb{R}$ with the Euclidean topology.

Exercise 3.38 Prove the following Intermediate Value Theorem. Let $f : X \to \mathbb{R}$ be a continuous function from a topological space X to the topological space $\mathbb{R}$ with the Euclidean topology. If X is connected, then for all $x, z \in X$ and $c \in \mathbb{R}$ with

$$f(x) < c < f(z),$$

there exists a $y \in X$ with

$$f(y) = c.$$

Exercise 3.39 Prove that if X and Y are connected (respectively path connected) topological spaces, then so is $X \times Y$.

Exercise 3.40 Prove that a quotient space of a connected space is connected.

3.5 Compactness

The reader is probably aware of the importance of closed intervals in $\mathbb{R}$ and, more generally, of closed and bounded subsets of $\mathbb{R}$ or $\mathbb{R}^n$. For instance, the Bolzano-Weierstrass Theorem states that every infinite sequence in a closed interval $[a, b]$ admits a convergent subsequence, or the theorem that a continuous function on a closed interval is uniformly continuous. It turns out that the crucial property of these subsets allowing for the mentioned results is a topological one, namely that these subsets are compact. The aim of this section is to present the notion of compactness and prove several results leading to the characterization of the closed and bounded subsets of $\mathbb{R}$ (under the Euclidean topology) as precisely the compact sets. The section concludes with the statement of the important result, known as Tychonoff's Theorem, that the product of any family of compact topological spaces is itself compact.

Definition 3.24 Let X be a topological space and $C \subseteq X$ a subset. An *open covering* of C is a collection $\{U_i\}_{i\in I}$ of open sets whose union contains the entire set C, i.e.

$$C \subseteq \bigcup_{i\in I} U_i.$$

The set C is said to be *compact* if given any open covering $\{U_i\}_{i\in I}$ of C, there exists a *finite subcovering* of C. That is, there exists finitely many sets $U_{i_1}, \ldots, U_{i_m}$ from the given covering such that

$$C \subseteq U_{i_1} \cap \cdots \cap U_{i_m}.$$

Notice that C may coincide with the set X, so we may speak of the space X itself being compact.

Example 3.19 Any finite set C in any topological space is compact. This is clear since any open covering of a finite set must admit a finite subcovering, using at most as many open sets from the covering as there are points in the set C.

Example 3.20 Any unbounded set $Y \subseteq \mathbb{R}$, when $\mathbb{R}$ is given the Euclidean topology, is not compact. To see that, consider the open covering

$$Y \subseteq \bigcup_{k=1}^{\infty} (-k, k).$$

Clearly, no finite subcovering suffices to cover Y precisely because Y was chosen to be unbounded. In other words, any compact set in $\mathbb{R}$ must be bounded. Similarly, an open interval (a, b) is not compact in $\mathbb{R}$ either, since the open covering

$$(a,b) \subseteq \bigcup_{k=1}^{\infty} (a + \frac{1}{k}, \infty)$$

admits no finite subcovering of (a, b), as is easy to verify.

Example 3.21 For any set X, if X is endowed with the indiscrete topology, then there are only two open sets (or sometimes less) in existence and so every subset $C \subseteq X$ is certainly compact. If, however, X is given the discrete topology, then every singleton set $\{x\}$ is open, and thus any subset $C \subseteq X$ admits the open covering

$$C = \bigcup_{x \in C} \{x\}$$

and thus is compact if, and only if, C is finite.

The next example is important enough to be stated as a theorem.

Theorem 3.15 *A closed interval* $[a, b]$ *in* $\mathbb{R}$ *with the Euclidean topology is compact.*

Proof Suppose that an open covering $\{U_i\}_{i \in I}$ of $[a, b]$ is given, and the goal is to prove a finite subcovering exists. Define the set S of all $a \leq s \leq b$ for which a finite subcovering of $[a, s]$ exists, and thus the goal now is to show that $b \in S$. Now, since $[a, a] = \{a\}$, which can certainly be covered by just one element from the given covering, it follows that $a \in S$, and thus S is not empty. By definition, S is bounded, and thus has a supremum t. We will show that $t = b$. Firstly, take some U_{i_0} from the given covering for which $t \in U_{i_0}$. As U_{i_0} is open, we have

$$t \in (t - \varepsilon, t + \varepsilon) \subseteq U_{i_0}$$

for some $\varepsilon > 0$, and since $t - \varepsilon < t$ it follows that $t - \varepsilon \in S$. There is then a finite subcovering of $[a, t - \varepsilon]$, and by adjoining U_{i_0} to it we obtain a finite subcovering of $[a, t]$, and thus $t \in S$, and thus a finite subcovering for $[a, t]$ exists. But then, if $t < b$, then the finite subcovering of $[a, t]$ is (by considering the set containing t) also a finite subcovering of $[0, t + \delta]$ for a suitable $\delta > 0$, contrary to the minimality of t. We thus conclude that $t = b$ and thus that one can extract a finite subcover of all of $[a, b]$. □

Generally speaking, a compact subset need not be closed, as the following example shows.

Example 3.22 Let X be an arbitrary set and consider the cofinite topology on it. The open sets (other than $\emptyset$) are those whose complements are finite, and thus we may identify the closed sets (other than X itself) as precisely the finite subsets. We now show that any subset $C \subseteq X$ is compact. Suppose an open covering $\{U_i\}_{i\in I}$ of C is given. To extract a finite subcovering, consider any non-empty member U_i of the covering, and note that since it is open it misses at most finitely many of the elements of C. Thus only finitely many other open sets from the given covering are required to cover C, and thus a finite subcovering exists. In particular then, if X is infinite, then not every compact subset of it is closed.

Lemma 3.2 *A closed subset C of a compact space X is compact.*

Proof Given an open covering of C, simply note that adding to it the set $X - C$, which is open since C is assumed closed, yields an open covering of X, which is compact. A finite subcovering of X thus must exist and discarding $X - C$ from it if necessary yields a finite subcovering of C, as required. □

The compact subsets of $\mathbb{R}$ with the Euclidean topology can be characterized as precisely the closed and bounded sets, as we proceed to show, starting with the following independently important result.

Proposition 3.12 *A compact subset C of a Hausdorff space X is closed.*

Proof If $C = X$, then C is closed, so we may assume $X - C$ is not empty, and we proceed to show that it is open. Let $x_0 \in X - C$ be an arbitrary point. For every $y \in C$, using the Hausdorff assumption on X, there exist disjoint open sets U_y and V_y such that $x_0 \in U_y$ and $y \in V_y$. Clearly, the collection $\{V_y\}_{y\in C}$ is an open covering of C, and since C is compact there exists a finite subcovering $V_{y_1}, \ldots, V_{y_k}$ such that

$$C \subseteq V_{y_1} \cup \cdots \cup V_{y_k}.$$

Let

$$U = U_{y_1} \cap \cdots \cap U_{y_k}$$

and note that U is open, that $x_0 \in U$, and that $U \subseteq C$. In other words, x_0 is an interior point of $X - C$ and since x_0 was an arbitrary point, every point of $X - C$ is interior, and thus, by Corollary 3.1, $X - C$ is open. □

Theorem 3.16 (Heine-Borel) *With respect to the Euclidean topology on $\mathbb{R}$, a subset $C \subseteq \mathbb{R}$ is compact if, and only if, C is closed and bounded.*

Proof Assume that $C \subseteq \mathbb{R}$ is compact subset. Since $\mathbb{R}$ is Hausdorff, it follows by Proposition 3.12 that C is closed while the fact that C must be bounded was already noted in Example 3.20. In the other direction, if C is bounded, then it is contained in some closed interval $[a, b]$, which by Theorem 3.15 is a compact set. If C is also closed then it is a closed subset of the compact set $[a, b]$ and thus, by Lemma 3.2, is itself compact. □

The next result is a generalization of the Bolzano–Weierstrass Theorem.

Theorem 3.17 *An infinite subset S of a compact topological space X has a cluster point in X.*

Proof Suppose that S does not have any cluster points in X. That means that for every $x \in X$ there exists an open set U_x with $x \in U_x$ and such that $U_x \cap S$ is either empty or the set $\{x\}$. Clearly, $\{U_x\}_{x \in X}$ is an open cover of X, and thus admits a finite subcover, say $U_{x_1}, \ldots, U_{x_n}$, and thus

$$S \subseteq U_{x_1} \cup \cdots \cup U_{x_n}.$$

But $U_{x_i} \cap S$ contains at most one element, and we thus conclude that $|S| \leq n$, so S is finite. Consequently, if S is infinite, then it must have at least one cluster point. □

We end this section by mentioning the following important result.

Theorem 3.18 (Tychonoff Theorem) *The product of any number (finite or infinite) of compact topological spaces is itself compact.*

The proof of this result is slightly beyond the scope of this book.

Exercises

Exercise 3.41 Show that a subset of a compact space may be compact but may also be non-compact. Similarly, show that a subset of a non-compact space may be compact but may also be non-compact.

Exercise 3.42 Prove, without appealing to Tychonoff's Theorem, that the product of any finite number of compact topological spaces is compact.

Exercise 3.43 Prove that if C is a non-empty compact subset of $\mathbb{R}$ with respect to the Euclidean topology, then C admits both a minimum and a maximum.

Exercise 3.44

1. Prove that if $f : X \to Y$ is a continuous function between topological spaces and $C \subseteq X$ is compact, then $f(C)$ is compact in Y.
2. Prove the following Extreme Value Theorem. Let $f : X \to \mathbb{R}$ be a continuous function from a topological space X to the topological space $\mathbb{R}$ with the Euclidean topology. If X is compact and non-empty, then f attains both a maximum and a minimum, i.e., there exist points $x_m, x_M \in X$ such that

$$f(x_m) \leq f(x) \leq f(x_M)$$

for all $x \in X$.
3. Deduce Weierstrass' Theorem: a continuous function $f : [a, b] \to \mathbb{R}$ attains both a minimum and a maximum.

Exercise 3.45 Prove that compactness is a topological invariant, namely if X and Y are homeomorphic topological spaces and one is compact, then so is the other.

Exercise 3.46 Prove that the coproduct of finitely many compact spaces is itself compact. Does the result remain valid for the coproduct of infinitely many compact spaces?

Exercise 3.47 Prove that a quotient space of a compact space is itself compact.

Exercise 3.48 Construct a topological space X with a non-empty proper subset which is both open and compact.

Exercise 3.49 Show that if X is an uncountable set endowed with the cocountable topology, then X is not compact. What are the compact subsets of it?

Exercise 3.50 A topological space X satisfies the *finite intersection property* if for any collection $\mathscr{F}$ of closed sets, if

$$F_1 \cap \cdots \cap F_m \neq \emptyset$$

for all $F_1, \ldots, F_m \in \mathscr{F}$, then

$$\bigcap_{F \in \mathscr{F}} F \neq \emptyset.$$

Prove that a topological space X has the finite intersection property if, and only if, X is compact.

Further Reading

For a perspective on some of the ideas that led to the birth of topology see [4]. For a comprehensive introduction to topology, including most of the results in this chapter and far more, the reader is referred to [3, 5]. For a remarkably unexpected usage of topology, illustrating its versatility, see Furstenberg's proof of the infinitude of primes [1]. For a collection of essays on various topological developments, including an essay devoted to topology and physics, see [2].

References

1. H. Furstenberg, On the infinitude of primes. Amer. Math. Mon. **62**, 355 (1955)
2. I.M. James (ed.), History of Topology (Amsterdam, North-Holland, 1999), p. 1056
3. J.R. Munkres, Topology: A First Course (Prentice-Hall Inc, Englewood Cliffs, 1975), p. 413
4. D.S. Richeson, *Euler's Gem, The Polyhedron formula and the birth of topology, First paperback printing* (Princeton University Press, Princeton, 2012)
5. T.B. Singh, Elements of Topology (CRC Press, Boca Raton, 2013), p. 530

Chapter 4
Metric Spaces

Abstract This chapter is concerned with the basic concepts of metric spaces and results that are most relevant to applications in Hilbert space theory. Assuming no prior knowledge of metric spaces, the definitions are given in detail and the relation to normed spaces made explicit early on. The main theorems proved are the Banach Fixed-Point Theorem, Baire's Theorem, the equivalence between compact sets and complete totally bounded sets, and an account of completion.

Keywords Metric space · Uniformly continuous function · Isometry · Banach fixed-point Theorem · Baire's Theorem · Cauchy condition · Metric completion · Totally bounded space · Metric topology · Compact metric space

The concept metric space was introduced by Maurice René Fréchet in 1906 in his Ph.D. dissertation devoted to functional analysis. The axioms given by Fréchet (which are almost identical to the ones we give below) form an abstraction of the notion of distance thus allowing for a unified treatment of numerous particular cases under a single formalism. The ubiquity of metric spaces in Mathematics and Physics is quite overwhelming but, naturally, this chapter is mainly concerned with examples and results most relevant in the context of this book.

Section 4.1 contains the definition of metric space and explores some examples, primarily of normed spaces as metric spaces. Section 4.2 is concerned with the topology induced by the distance function, the associated concept of convergence, and some basic topological observations regarding the induced topology. Section 4.3 considers two types of structure preservation for mappings between metric spaces, namely non-expanding mappings and uniformly continuous ones. Section 4.4 is concerned with the concept of completeness in metric spaces, the Banach Fixed-Point Theorem and Baire's Theorem, and Cantor's metric completion process by means of Cauchy sequences. Finally, Sect. 4.5 examines compactness for metric spaces and establishes the fact that a metric space is compact precisely when it is complete and totally bounded.

© Springer International Publishing Switzerland 2015

C. Alabiso and I. Weiss, *A Primer on Hilbert Space Theory*,
UNITEXT for Physics, DOI 10.1007/978-3-319-03713-4_4

4.1 Metric Spaces—Definition and Examples

Our intuitive understanding of the behaviour of distance, in the plane for instance, is clearly represented in the definition of metric space that we give below. Of particular importance is the simple fact that any normed space has a natural metric structure associated with it (though most metric spaces certainly do not arise in this manner). Consequently, the normed spaces encountered in Chap. 2 are all revisited here so that we may examine some of their metric properties.

Before we proceed with the definition and with a list of relevant examples we note that for the general study of metric spaces there are technical advantages in allowing infinite distance between points. We thus first adjoin the symbol ∞ as an extended real number and denote by $\mathbb{R}_+ = \{x \in \mathbb{R} \mid x \geq 0\} \cup \{\infty\}$ the set of *extended non-negative real numbers*.

Remark 4.1 We accept that $x < \infty$ for all real x (and we also accept that $\infty \leq \infty$). Addition is extended to include ∞ by setting $x + \infty = \infty + x = \infty$ for all extended real $x \geq 0$.

Definition 4.1 Let X be a set. A function $d : X \times X \to \mathbb{R}_+$ is said to be a *metric* or a *distance function* on X if, for all $x, y, z \in X$, the following properties are satisfied.

1. Distance is symmetric, i.e., $d(x, y) = d(y, x)$.
2. The triangle inequality is satisfied, i.e., $d(x, z) \leq d(x, y) + d(y, z)$.
3. Distance separates points, i.e., $d(x, y) = 0$ if, and only if, $x = y$.

The pair (X, d) is then called a *metric space* and $d(x, y)$ the *distance* between x and y. Quite often we refer to the metric space X without explicitly mentioning the distance function d. Moreover, when two (or more) metric spaces are considered, we often overload the symbol d to stand for the metric function in each space, relying on semantics to resolve any confusion. If needed, we may refer to the metric function of X by d_X.

Remark 4.2 There is quite a lot of flexibility in the choice of axioms above, giving rise to several structures that may resemble metric spaces to varying degrees. For instance, as noted above, it is possible that $d(x, y) = \infty$, that is we allow the distance between points to be infinity. Some authors do not permit the distance function to attain the value ∞. In some sense, the difference between these two possibilities is largely cosmetic. It is also possible to omit the separation condition, giving rise to what are known as *semimetric* spaces, and again, their theory is quite similar to that of metric spaces. If instead one neglects symmetry, giving rise to what are known as *quasimetric* spaces, the resulting theory is in many respects strikingly different than that of metric spaces.

Observing that all of the axioms of metric space are universally quantified, it follows immediately that if $S \subseteq X$ is a subset of a metric space X, then the restriction of the distance function $d : X \times X \to \mathbb{R}_+$ to the subset $S \times S$ is a distance function on S.

Definition 4.2 (*Metric Subspace*) For a subset S of a metric space X the pair (S, d_S), where d_S is the restriction of d to $S \times S$, is called a *metric subspace* (or simply *subspace* if the metric context is clear) of X.

There are various ways for a function f between metric spaces to interact with the metric structures, namely to preserve the distances to various degrees. The strictest preservation of distance gives rise to the following definition.

Definition 4.3 A function $f : X \to Y$ between metric spaces is said to be an *isometry* if

$$d(f(x_1), f(x_2)) = d(x_1, x_2)$$

for all $x_1, x_2 \in X$. If f is also bijective, then f is said to be an *isometric isomorphism* or a *global isometry*. If there exists an isometric isomorphism between X and Y, then we write $X \cong Y$ and we say that X and Y are *isometric*.

A property of metric spaces is said to be a *metric invariant* if whenever it holds for a metric space X it also holds for any other metric space isometric to X.

The following theorem is a rich source of metric spaces.

Theorem 4.1 *If V is a normed space, then defining $d : V \times V \to \mathbb{R}_+$ by*

$$d(x, y) = \|x - y\|$$

endows V with the structure of a metric space.

Proof Note first that the codomain of d is indeed $\mathbb{R}_+$ since the norm is always non-negative (in this case ∞ is never attained as the distance between points). For all vectors $x, y, z \in V$, symmetry follows by

$$d(x, y) = \|x - y\| = \|(-1) \cdot (y - x)\| = |-1| \cdot \|y - x\| = \|y - x\| = d(y, x),$$

while separation and the triangle inequality are just restatements of the conditions of positivity and the triangle inequality in the definition of normed space (refer to Definition 2.15, and with the aid of Proposition 2.12). □

Example 4.1 Each of the normed spaces (and in particular the inner product spaces) introduced in Chap. 2 gives rise to a metric space, and we obtain the following examples of metric spaces.

The space $\mathbb{R}^n$, $n \geq 1$, with distance given by

$$d(x, y) = \left(\sum_{k=1}^{n} (x_k - y_k)^2 \right)^{1/2},$$

known as the *Euclidean metric* on $\mathbb{R}^n$. The space $\mathbb{C}^n$, $n \geq 1$, with distance given by

$$d(x, y) = \left(\sum_{k=1}^{n} |x_k - y_k|^2 \right)^{1/2},$$

where we may immediately note that $\mathbb{C}^n$ is isometric to $\mathbb{R}^{2n}$ with the Euclidean metric, a global isometry $f : \mathbb{C}^n \to \mathbb{R}^{2n}$ is given by

$$f(x_1 + y_1 \cdot i, \ldots, x_n + y_n \cdot i) = (x_1, y_1, x_2, y_2, \ldots, x_n, y_n).$$

For all $1 \leq p < \infty$, every pre-L_p space, say of continuous real-valued functions on the closed interval $[a, b]$, is a metric space of functions with distance given by

$$d(x, y) = \left(\int_a^b dt\ |x(t) - y(t)|^p \right)^{1/p}.$$

In particular, $C([a, b], \mathbb{R})$ supports infinitely many different metric space structures, one for each $1 \leq p < \infty$. The case $p = \infty$, i.e., with the L_∞ norm, gives rise to yet another distance function, namely

$$d(x, y) = \max_{a \leq t \leq b} |x(t) - y(t)|.$$

Similarly, the space $C(I, \mathbb{C})$ of continuous complex-valued functions on a closed interval $I = [a, b]$ supports an entire family of metric space structures. For each $1 \leq p < \infty$ the space ℓ_p of absolutely p-power summable sequences of either real or complex numbers with distance given by

$$d(x, y) = \left(\sum_{k=1}^{\infty} |x_k - y_k|^p \right)^{1/p}$$

is yet another family of metric spaces. Of course the case $p = \infty$, i.e., the space ℓ_∞ of bounded sequences with distance given by

$$d(x, y) = \sup_{k \geq 1} |x_k - y_k|$$

is a metric space as well. Beware here not to confuse the value ∞ for the parameter p with a limiting process.

$\mathbb{R}^n$ and $\mathbb{C}^n$ support other metric structures, different than those presented above, as the following example shows.

Example 4.2 If we consider $\mathbb{R}^n$ as a subset of ℓ_p, with $1 \leq p \leq \infty$, by identifying $x = (x_1, \ldots, x_n)$ with $(x_1, \ldots, x_n, 0, 0, 0, \ldots)$, then $\mathbb{R}^n$ inherits the metric of ℓ_p and becomes a metric subspace of ℓ_p. In more detail, if $p = \infty$ then

$$d(x, y) = \max_{1 \le k \le n} |x_k - y_k|$$

and otherwise

$$d(x, y) = \left(\sum_{k=1}^{n} |x_k - y_k|^p \right)^{1/p}.$$

If $n > 1$ and $p \neq 2$, then each of these metrics is different than the Euclidean metric. A similar observation gives rise to a family of metric structures on $\mathbb{C}^n$.

The following examples illustrate the subtleties that may arise with general metric spaces.

Example 4.3 Consider $\mathbb{R}^2$ with the Euclidean metric, i.e., the plane with its ordinary Euclidean metric structure. The unit circle $S = \{x \in \mathbb{R}^2 \mid \|x\| = 1\}$ has two natural metric structures on it. As a subspace of $\mathbb{R}^2$, the distance between two antipodal points in S is 2 (the length of the shortest straight line connecting the two points where the line is allowed to pass freely in the ambient space $\mathbb{R}^2$). On the other hand, it is possible to endow the unit circle S with the so called *intrinsic metric*, the one where the distance between two points is the length of the shorter of the two arcs connecting them. With that metric the distance between antipodal points is π (the length of a shortest geodesic connecting the two points, that is a "straight" line connecting the points where the line is not allowed to step outside of S and into the ambient space).

Other scenarios, perhaps more abstract, are also possible. For instance, given any set X, defining $d(x, x) = 0$ for all $x \in X$ and $d(x, y) = 1$ for all $x, y \in X$ with $x \neq y$ turns X into a metric space. In this metric space the distances between any two distinct points is 1. Thus, if X is infinite, then X is not isometric to any subspace of the Euclidean space $\mathbb{R}^n$, for any $n \geq 1$. Notice however that defining $d(x, y) = 0$ for all $x, y \in X$ fails to be a metric space (unless X has at most one element) since separation fails. It is however easily seen to be an example of a semimetric space.

Example 4.4 The set $\mathbb{R}_+$ can be given the metric structure

$$d(x, y) = \begin{cases} |x - y| & \text{if } x \neq \infty \neq y \\ 0 & \text{if } x = y = \infty \\ \infty & \text{otherwise.} \end{cases}$$

The metric axioms are easily verified. In fact, we will simply write $d(x, y) = |x - y|$ since the algebraic properties of this metric significantly resemble those of $|x - y|$ for real x, y.

Finally, we present one of many ways to endow the cartesian product $X \times Y$ of two metric spaces with a metric structure.

Definition 4.4 (*The Additive Metric Product*) For metric spaces X and Y the function $d : (X \times Y) \times (X \times Y) \to \mathbb{R}_+$ given by

$$d((x, y), (x', y')) = d_X(x, x') + d_Y(y, y')$$

is easily seen to be a distance function (as the reader may verify) and the metric space $(X \times Y, d)$ is called the *additive metric product* of X and Y.

Exercises

Exercise 4.1 Let $f : [0, \infty) \to [0, \infty)$ be a concave, strictly monotonically increasing function satisfying $f(0) = 0$. Suppose that (X, d) is a metric space and that d only attains finite values. Prove that $(X, f \circ d)$ is a metric space.

Exercise 4.2 Let (X, d_X) and (Y, d_Y) be two disjoint (i.e., $X \cap Y = \emptyset$) metric spaces and let $Z = X \cup Y$. Prove that $d : Z \times Z \to \mathbb{R}_+$ given by

$$d(z_1, z_2) = \begin{cases} d_X(z_1, z_2) & \text{if } z_1, z_2 \in X \\ d_Y(z_1, z_2) & \text{if } z_1, z_2 \in Y \\ \infty & \text{otherwise} \end{cases}$$

is a metric on Z. The metric space (Z, d) is then called the *coproduct* of X and Y. Generalize this construction to give the coproduct of any number (finite or infinite) of metric spaces.

Exercise 4.3 Let X be a metric space. Define the relation $\sim$ on X by $x \sim y$ precisely when $d(x, y) < \infty$. Prove that $\sim$ is an equivalence relation. The equivalence class $[x]$, for any point $x \in X$, is called the *galaxy* of x. Show that each galaxy, as a metric subspace of X, has all distances finite and then show that X is the coproduct (see previous exercise) of its galaxies.

Exercise 4.4 Suppose that (X, d) is a semimetric space and define the relation $\sim$ on X by $x \sim y$ precisely when $d(x, y) = 0$. Prove that $\sim$ is an equivalence relation on X and that $d([x], [y]) = d(x, y)$ on the quotient set $X/\sim$ is well-defined and endows it with the structure of a metric space.

Exercise 4.5 Given any two metric spaces (X, d_X) and (Y, d_Y), define the function $d_E : (X \times Y) \times (X \times Y) \to \mathbb{R}_+$ by

$$d_E((x_1, y_1), (x_2, y_2)) = \sqrt{d_X(x_1, x_2)^2 + d_Y(y_1, y_2)^2}.$$

Prove that $(X \times Y, d_E)$ is a metric space, called the *Euclidean metric product* of X and Y. Generalize this construction to obtain the Euclidean metric product of any finite number of metric spaces.

Exercise 4.6 Prove that every isometry is injective.

Exercise 4.7 For all metric spaces X, Y, and Z prove that

1. $X \cong X$

2. If $X \cong Y$, then $Y \cong X$
3. If $X \cong Y$ and $Y \cong Z$, then $X \cong Z$.

Exercise 4.8 Prove that the induced metric in a normed space V satisfies the equality $d(x+z, y+z) = d(x, y)$ for all vectors $x, y, z \in V$. Namely, the induced metric in a normed space V is *translation invariant*.

Exercise 4.9 Consider the set R of all Riemann integrable functions $f : [a, b] \to \mathbb{R}$ and define $d : R \times R \to \mathbb{R}_+$ by $d(f, g) = \int_a^b dt\, |f(t) - g(t)|$. Is (R, d) a metric space?

Exercise 4.10 A *mid-point* for two distinct points x, z in a metric space X is a point $y \in X$ such that $d(x, y) = d(y, z) = d(x, z)/2$. Give examples of:

- A metric space where every two distinct points admit a unique mid-point.
- A metric space where every two distinct points admit a mid-point, and some admit two distinct mid-points.
- A metric space where no two distinct points admit a mid-point.

4.2 Topology and Convergence in a Metric Space

This section is concerned with the topological structure of a metric space. Every metric space has a topology associated with it which is determined by the distance function. A metric space is thus automatically a topological space and thus all of the concepts of topological spaces, including convergence, are meaningful in a metric space. After describing this so called induced topology we examine the concept of convergence in the metric spaces introduced above. It should be noted immediately that different metrics on the same set may induce the same topology. A topological space induced by a metric is called a metrizable space and we study some of the properties of such spaces.

As motivation for the induced topology we follow a similar route to the one we took in Sect. 3.1 where the concept of topology was motivated as a means to capture continuity while avoiding arbitrary properties of the distance function.

Before we proceed we note that the familiar notion of continuity of a function applies (formally verbatim) to arbitrary metric spaces.

Definition 4.5 A function $f : X \to Y$ between metric spaces is *continuous* if for all $x \in X$ and $\varepsilon > 0$ there exists a $\delta > 0$ such that

$$d(x, x') < \delta \implies d(f(x), f(x')) < \varepsilon$$

for all $x' \in X$.

4.2.1 The Induced Topology

The topology induced by a metric is given in terms of what are known as open balls, which are shown below to form a basis for a topology. It should be made clear at once that the resulting topology does not carry all of the metric information. In fact, it carries very little of it. There may be many radically different distance functions on a single set, all of which induce the same topology. Moreover, not every topology is necessarily induced by a metric. Thus, replacing a metric by the induced topology looses a lot of information and misses many topological spaces. However, it is often convenient to be able to forget information, especially when it is not really needed for the problem at hand.

Definition 4.6 Let X be a metric space, $x \in X$, and $\varepsilon > 0$. The *open ball* with *centre* x and *radius* ε is the set

$$B_\varepsilon(x) = \{y \in X \mid d(x, y) < \varepsilon\}.$$

Example 4.5 Consider the Euclidean metric on $\mathbb{R}^n$. The open ball $B_\varepsilon(x)$ in $\mathbb{R}$ is the open interval $(x - \varepsilon, x + \varepsilon)$, while in $\mathbb{R}^2$ it is the interior of the circle with centre x and radius ε, and in $\mathbb{R}^3$ it is the interior of a ball with centre x and radius ε.

Obviously, in a general metric space an open ball need not look anything like a ball. Figure 4.1 shows the loci of points in $\mathbb{R}^2$ satisfying $\|x\| = 1$ under the ℓ_p norm for different values of p. Note that the case $p = 1/2$ is not at all a metric and is just added here for the sake of completeness.

We now show that the open balls in a metric space X form a basis for a topology.

Theorem 4.2 *The collection* $\{B_\varepsilon(x) \mid x \in X, \varepsilon > 0\}$ *of all open balls in a metric space* X *satisfies the conditions of Theorem* 3.10.

Proof Since $x \in B_\varepsilon(x)$ for all $x \in X$ and $\varepsilon > 0$, it follows at once that the open balls cover all of X. Next, suppose that $z \in B_{\varepsilon_1}(x) \cap B_{\varepsilon_2}(y)$, and we will demonstrate that $z \in B_\varepsilon(z) \subseteq B_{\varepsilon_1}(x) \cap B_{\varepsilon_2}(y)$ for $\varepsilon = \min\{\varepsilon_1 - d(x, z), \varepsilon_2 - d(y, z)\}$. Indeed, if $d(w, z) < \varepsilon$, then $d(w, x) \leq d(w, z) + d(z, x) < (\varepsilon_1 - d(x, z)) + d(z, x) = \varepsilon_1$, and so $w \in B_{\varepsilon_1}(x)$, and therefore $B_\varepsilon(z) \subseteq B_{\varepsilon_1}(x)$. Similarly, $B_\varepsilon(z) \subseteq B_{\varepsilon_2}(y)$. Noting that $\varepsilon > 0$, the proof is complete. □

The following definition is now justified.

Definition 4.7 Given a metric space (X, d), the *induced topology* on the set X is the topology generated by the open balls in X. In particular, all open balls in X are open sets, and a subset $U \subseteq X$ is open in the induced topology precisely when for all $x \in U$ there exists an $\varepsilon > 0$ such that $B_\varepsilon(x) \subseteq U$. Any topological space (X, τ) for which there exists a metric d on X inducing the given topology τ is called a *metrizable space*.

Remark 4.3 Unless otherwise stated, every metric space is silently endowed with the induced topology. Consequently, every metric space X is also a topological space.

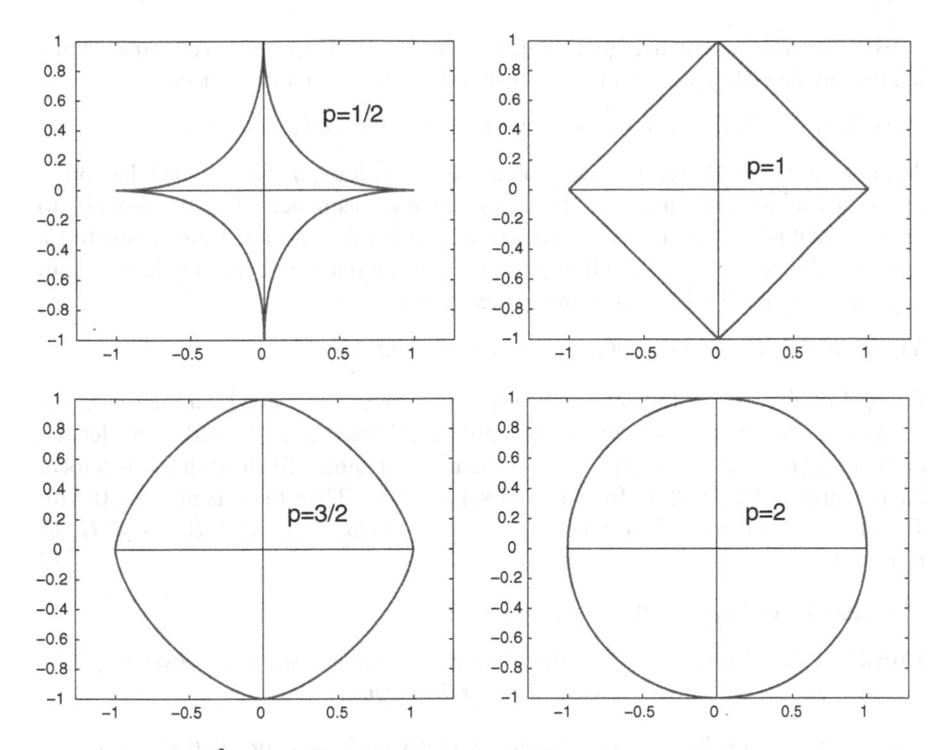

Fig. 4.1 Points in $\mathbb{R}^2$ satisfying $\|x\| = 1$ with $\|x\|^p = \sum |x_k|^p$, for different values of p

Example 4.6 It is straightforward to see that the Euclidean distance $d(x, y) = |x-y|$ on $\mathbb{R}$ induces the Euclidean topology on $\mathbb{R}$. One may also readily confirm that the discrete topology on a set X is induced by the distance function $d : X \times X \to \mathbb{R}_+$ given by $d(x, y) = 1$ if $x \neq y$ and $d(x, x) = 0$, thus any discrete topological space is metrizable. Examples of non-metrizable spaces are easily constructed once a few basic facts about metrizable spaces are observed, as done below.

Theorem 4.3 *A function $f : X \to Y$ between metric spaces is continuous in the sense of Definition 4.5 if, and only if, it is continuous with respect to the induced topologies.*

The proof is formally identical to the proof of Theorem 3.2 and is left for the reader. We now turn to examine some general properties of metrizable spaces.

Theorem 4.4 *Every metrizable space is Hausdorff.*

Proof Let X be a topological space whose topology is induced by a metric d. Given distinct points $x, y \in X$ let $\varepsilon = d(x, y)/2$, and notice that $\varepsilon > 0$. We claim that the open balls $B_\varepsilon(x)$ and $B_\varepsilon(y)$ separate x and y. Indeed, if $z \in B_\varepsilon(x) \cap B_\varepsilon(y)$, then we would have $d(x, y) \leq d(x, z) + d(z, y) < 2\varepsilon = d(x, y)$, which is absurd. □

By Theorem 3.12, convergent sequences in Hausdorff spaces have unique limits, leading to the following familiar result stated for general metric spaces.

Corollary 4.1 *In a metric space, a convergent sequence has a unique limit.*

Apart from being Hausdorff, every metrizable topology, as we will see, has other pleasant features influencing the behaviour of, e.g., sequences. It is worth-while to notice that it is the interaction of a metric space with $\mathbb{R}_+$ (via the distance function), which causes some of the familiar properties of the real numbers to reflect back to any metric space. This is shown in the next result.

Theorem 4.5 *Every metrizable space is first countable.*

Proof Let X be a topological space whose topology is induced by a metric d, and let $x \in X$. We need to exhibit a countable local basis at x. Indeed, consider the collection $\{B_{1/n}(x) \mid n \in \mathbb{N}\}$, which is clearly countable. To show that it is a local basis, suppose that $x \in U$ for an open set $U \subseteq X$. Then there is an $\varepsilon > 0$ with $B_\varepsilon(x) \subseteq U$, and thus if we take any $n > 1/\varepsilon$, then $B_{1/n}(x) \subseteq B_\varepsilon(x) \subseteq U$, as required. □

Theorem 3.9 yields the following corollaries.

Corollary 4.2 *A point x in a metric space X is a cluster point of a subset $S \subseteq X$ if, and only if, x is the limit of a sequence of points from S.*

Corollary 4.3 *A subset F of a metrizable space X is closed if, and only if, F contains all limit points of sequences in F.*

Corollary 4.4 (Heine's Continuity Criterion) *A function $f : X \to Y$ between two metrizable spaces is continuous if, and only if,*

$$x_m \to x_0 \implies f(x_m) \to f(x_0)$$

for all sequences $\{x_m\}_{m \geq 1}$ in X.

By Theorem 3.8, a second countable space is separable. We now show that for metric spaces the converse holds as well.

Theorem 4.6 *A separable metrizable space X is second countable.*

Proof Let d be a metric inducing the topology on X. We are given the existence of a countable and dense set $D = \{x_k\}_{k \in \mathbb{N}}$ in X and we need to exhibit a countable basis for X. In the proof of Theorem 4.5 we saw that the collection

$$\mathscr{B}_k = \{B_{1/n}(x_k) \mid n \in \mathbb{N}\}$$

is a local basis at x_k. We now show that

$$\mathscr{B} = \bigcup_{k \in \mathbb{N}} \mathscr{B}_k$$

is a basis for the topology of X, noticing that the result then follows since B, being a countable union of countable sets, is itself countable (see Sect. 1.2.13 of the Preliminaries).

Let $U \subseteq X$ be an open subset, and choose $x \in U$. There is then an $\varepsilon > 0$ such that $B_{2\varepsilon}(x) \subseteq U$. Choose $n \in \mathbb{N}$ such that $1/n < \varepsilon$. Since D is dense and $B_{1/n}(x)$ is open, it follows that $D \cap B_{1/n}(x)$ is not empty, and thus there is some m such that $x_m \in B_{1/n}(x)$, in other words, $d(x, x_m) < 1/n$. Now, the choice of n guarantees that $B_{1/n}(x_m) \subseteq B_{2\varepsilon}(x) \subseteq U$, while the choice of x_m guarantees that $x \in B_{1/n}(x_m)$. Noticing that $B_{1/n}(x_m) \in \mathscr{B}_m$, the above establishes that for every open set U and $x \in U$, there exists a $B \in \mathscr{B}$ with $x \in B \subseteq U$, showing that $\mathscr{B}$ is a basis for the topology on X. □

4.2.2 Convergence in Metric Spaces

As remarked above, every metric space is in particular a topological space and thus has an intrinsic notion of convergence which we now turn to investigate for the main examples of metric spaces given above.

Proposition 4.1 *A sequence $\{x_m\}_{m\geq 1}$ in a metric space X converges to $x_0 \in X$ if, and only if, $d(x_m, x_0) \to 0$ in $\mathbb{R}$ in the usual sense of convergence.*

Proof Suppose $x_m \to x_0$ in the topological sense. Given $\varepsilon > 0$ consider the open ball $B_\varepsilon(x_0)$, which is open in the induced topology. That means that there exists an $N \in \mathbb{N}$ such that $x_m \in B_\varepsilon(x_0)$ for all $m > N$. In other words, $d(x_m, x_0) < \varepsilon$ for all $m > N$, showing that $d(x_m, x_0) \to 0$. The converse is left for the reader. □

Example 4.7 Consider $\mathbb{R}^n$ with the Euclidean metric

$$d(x, y) = \left(\sum_{k=1}^{n} (x_k - y_k)^2\right)^{1/2}.$$

For a sequence $\{x^{(m)}\}_{m\geq 1}$ of elements in $\mathbb{R}^n$ and a point $x \in \mathbb{R}^n$, the inequalities

$$|x_k^{(m)} - x_k| \leq d(x^{(m)}, x),$$

valid for all $1 \leq k \leq n$, show that $x^{(m)} \to x$ implies $x_k^{(m)} \to x_k$, for each component. The converse is also true, (since there are only finitely many components), and thus convergence in the Euclidean metric in $\mathbb{R}^n$ is equivalent to the usual convergence of the components in $\mathbb{R}$.

Example 4.8 Consider the space ℓ_p, $p \geq 1$ (see Sect. 2.5.4). Much as in the previous example, convergence in the ℓ_p norm implies the convergence of each of the components. However, as each vector now has infinitely many components, the converse does not necessarily hold. To see that, take the sequence $\{x^{(m)}\}_{m\in\mathbb{N}}$ where

$$x_k^{(m)} = \begin{cases} 1 & \text{if } m = k, \\ 0 & \text{if } m \neq k. \end{cases}$$

Each sequence $\{x_k^{(m)}\}_{m\in\mathbb{N}}$ is eventually constantly 0, and thus the only candidate for the limit of $\{x^{(m)}\}_{m\in\mathbb{N}}$ is the 0 vector. However,

$$d(x^{(m)}, 0) = \left(\sum_{i=1}^{\infty} |x_k^{(m)}|^p \right)^{1/p} = 1,$$

and thus $x^{(m)} \not\to 0$ in the metric sense, and therefore $\{x^{(m)}\}_{m\geq 1}$ does not converge at all in ℓ_p. The same sequence exhibits similar behaviour in ℓ_∞ too.

Example 4.9 Consider the space $C(I, \mathbb{R})$ of continuous real-valued functions on the closed interval $I = [a, b]$, with the metric induced by the L_∞ norm, i.e., for a sequence of functions $\{x_m\}_{m\geq 1}$ and a function x_0:

$$x_m \to x_0 \qquad \Longleftrightarrow \qquad \max_{t\in[a,b]} |x_m(t) - x_0(t)| \to 0.$$

Thus, for any fixed $\varepsilon > 0$ there exists an $N \in \mathbb{N}$ such that $|x_m(t) - x_0(t)| < \varepsilon$ for all $m > N$ and for all $t \in [a, b]$. In other words, x_m converges to x_0 uniformly on $[a, b]$. Since the converse is also true, as is easily seen by traversing the arguments backwards, we conclude that convergence in the L_∞ norm is nothing but uniform convergence of functions on the interval $[a, b]$.

Example 4.10 For $1 \leq p < \infty$ consider the pre-L_p space $C(I, \mathbb{R})$, with $I = [a, b]$, i.e., with distance given by

$$d(x, y) = \left(\int_a^b dt\, |(x(t) - y(t)|^p \right)^{1/p}.$$

Convergence with respect to this metric is called L_p convergence. It is well-known, from elementary considerations of the Riemann integral, that if $x_m \to x_0$ uniformly, then x_m also converges to x_0 in the L_p norm. The converse is not generally true, as we exemplify for the L_2 norm. Consider the following sequence in $C([0, 1], \mathbb{R})$:

$$x_m(t) = \begin{cases} mt & \text{if } 0 \leq t \leq \frac{1}{m}, \\ 1 & \text{if } \frac{1}{m} < t \leq 1. \end{cases}$$

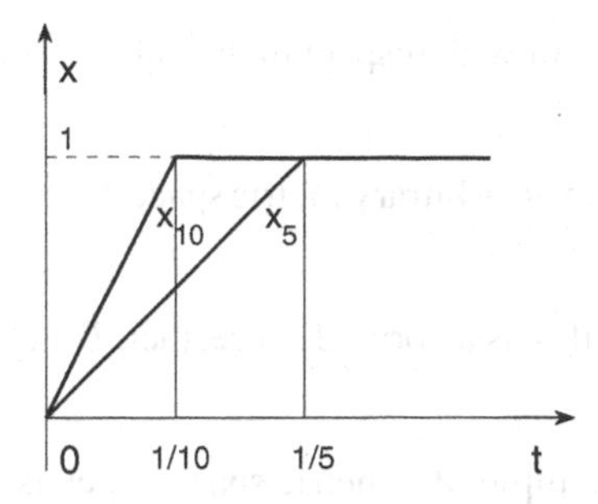

The computation

$$d(x_m, x)^2 = \int_0^1 dt\, \left(x_m(t) - 1\right)^2 = \int_0^{1/m} dt\, (mt - 1)^2 = \frac{1}{3m} \to 0$$

shows that x_m converges in L_2 to the constant function $x(t) = 1$. However, point-wise, $x_m(t)$ converges to the discontinuous function $\tilde{x}(t) = 1$ for $0 < t \leq 1$ and $\tilde{x}(0) = 0$. Since the uniform limit of continuous functions must be continuous, it follows that $\{x_m\}_{m\geq 1}$ does not converge with respect to the L_∞ norm.

Example 4.11 In fact, L_p convergence need not even imply point wise convergence at any t in the domain. Indeed, consider $C([0, 1], \mathbb{R})$ and, for simplicity, the L_1 norm. Let I_m be the following sequence of intervals.The first interval is $I_0 = [0, 1]$, the entire segment. The next two intervals are $I_1 = [0, 1/2]$ and $I_2 = [1/2, 1]$. The next three intervals are obtained by subdividing $[0, 1]$ into three equal segments, the next four intervals are obtained by subdividing $[0, 1]$ into four equal segments, and so on. Consider for each m the indicator function $x_m = \chi_{I_m}$. These functions are not continuous, but let us ignore this for a while. It is clear that, with respect to the L_1 norm, x_m converges to the constant function 0, since the integral $\int_0^1 dt\, \chi_m$ is the length of the segment that was considered in the m-th stage, and the segments' sizes shrink to 0. However, for any $0 \leq t \leq 1$, the sequence $\{\chi_m(t)\}_{m\geq 1}$ contains a subsequence which is constantly 1 and a subsequence which is constantly 0, and thus does not converge. This example would thus serve our purpose, illustrating that L_1 convergence need not imply point wise convergence at any point, except that the functions are not continuous. This can easily be remedied, with minor changes in the details above, by sacrificing a little bit of the segment used to define x_m in order to add two linear segments to the graph of χ_m so as to obtain the graph of a continuous function x_m. This can be done in such a way as to only slightly affect the integrals so that the L_1 convergence is unaltered, while retaining the point-wise non-convergence.

Exercises

Exercise 4.11 Let X be a metric space, $x \in X$, and $\varepsilon > 0$. The *closed ball* with centre x and radius ε is the set $\overline{B}_\varepsilon(x) = \{y \in X \mid d(x, y) \leq \varepsilon\}$. Prove that the closure of the open ball $B_\varepsilon(x)$ with respect to the induced topology is the closed ball $\overline{B}_\varepsilon(x)$.

Exercise 4.12 Show that in an arbitrary metric space X, the equality $B_\varepsilon(x) = B_\delta(y)$ need not imply $x = y$ nor $\varepsilon = \delta$.

Exercise 4.13 Prove that if X is a normed space, then $B_\varepsilon(x) = B_\delta(y)$ implies $x = y$ and $\varepsilon = \delta$.

Exercise 4.14 Give an example of a metric space which is not second countable.

Exercise 4.15 Prove Theorem 4.3.

Exercise 4.16 Let d_1 and d_2 be two distance functions on a set X such that there exists real numbers $\alpha, \beta > 0$ with

$$\alpha d_1(x, x') \leq d_2(x, x') \leq \beta d_1(x, x')$$

for all $x, x' \in X$. Prove that d_1 and d_2 induce the same topology on X.

Exercise 4.17 When is the cofinite topology on a set X metrizable?

Exercise 4.18 Prove that the coproduct of two metrizable spaces is metrizable. How far does this result generalize, e.g., considering more than two spaces?

Exercise 4.19 Prove that the product of two metrizable spaces is metrizable. How far does this result generalize, e.g., considering more than two spaces?

Exercise 4.20 Find a metrizable space and a quotient of it which is not metrizable.

4.3 Non-Expanding Functions and Uniform Continuity

Given two metric spaces X and Y, the rich structure embodied in the metrics can be preserved to various degrees by functions $f : X \to Y$. In a sense the weakest preservation of structure one can impose is for f to be continuous while the strongest is for f to be an isometry. In this section we look at two intermediate conditions for structure preservation, namely non-expansion and uniform continuity.

Definition 4.8 A function $f : X \to Y$ between metric spaces is *non-expanding* if

$$d(f(x), f(x')) \leq d(x, x')$$

for all $x, x' \in X$.

An important observation is that in any metric space (X, d) the distance function $d : X \times X \to \mathbb{R}_+$ is non-expanding.

Lemma 4.1 *Let X be a metric space and consider $X \times X$ with the additive metric product structure (Definition 4.4), and $\mathbb{R}_+$ with the metric structure described in Example 4.4. Then the distance function $d : X \times X \to \mathbb{R}_+$ is non-expanding.*

Proof Given $(x, x'), (y, y') \in X \times X$ we may assume without loss of generality that $d_X(x, x') \geq d_X(y, y')$. Let us further assume that both $d_X(x, x')$ and $d_X(y, y')$ are finite. It then follows that

$$d_{\mathbb{R}_+}(d(x, x'), d(y, y')) = |d_X(x, x') - d_X(y, y')| = d_X(x, x') - d_X(y, y').$$

Since

$$d_{X \times X}((x, x'), (y, y')) = d_X(x, y) + d_X(x', y')$$

we need to show that

$$d_X(x, x') \leq d_X(x, y) + d_X(y, y') + d_X(y', x')$$

which is indeed the case by the triangle inequality. Treating the cases where either $d_X(x, x') = \infty$ or $d_X(y, y') = \infty$ is similar. □

Non-expansion is clearly a rather strong property, quite close in strength to being an isometry. Uniform continuity, which we now introduce, is closer to the other end of the spectrum of structure preservation and, in a sense, is only slightly stronger than continuity.

Definition 4.9 A function $f : X \to Y$ between metric spaces is *uniformly continuous* if for every $\varepsilon > 0$ there is a $\delta > 0$ such that

$$d(x, x') < \delta \implies d(f(x), f(x')) < \varepsilon$$

for all $x, x' \in X$.

It is obvious that if $f : X \to Y$ is non-expanding, then it is uniformly continuous, and that uniform continuity implies continuity. Neither of the converse implications is generally true. However, the reader is familiar with the fact that continuity on a closed interval implies uniform continuity. The crucial property here is that a closed interval is compact. Indeed, we now show that if X is compact, then continuity does imply uniform continuity. The proof we present relies on the topological concept known as the Lebesgue number of a covering, requiring the following preparatory concept which is part of the standard vocabulary in metric space theory.

Definition 4.10 Let X be a metric space and $S \subseteq X$ a subset. The *diameter* of S is the extended real number

$$\text{diam}(S) = \sup_{x,y \in S} \{d(x, y)\}$$

(which may be ∞).

For instance, if $S = \{x, y\}$, then $\operatorname{diam}(S) = d(x, y)$. In $\mathbb{R}^n$ with the Euclidean metric $\operatorname{diam}(B_\varepsilon(x)) = 2\varepsilon$, for all $x \in \mathbb{R}^n$ and $\varepsilon > 0$. More generally, in an arbitrary metric space $\operatorname{diam}(B_\varepsilon(x)) \leq 2\varepsilon$ always holds and equality may fail.

Lemma 4.2 (Lebesgue Number Lemma) *Let X be a compact metric space. If $\{U_i\}_{i\in I}$ is an arbitrary open covering of X, then there exists a $\delta > 0$ such that for all subsets $S \subseteq X$ with $\operatorname{diam}(S) < \delta$ there is an $i \in I$ with $S \subseteq U_i$.*

Proof For each $x \in X$ there is an $i_x \in I$ with $x \in U_{i_x}$, and since U_{i_x} is open there is an $\varepsilon_x > 0$ such that $B_{2\varepsilon_x}(x) \subseteq U_{i_x}$. Clearly, the collection $\{B_{\varepsilon_x}(x)\}_{x\in X}$ covers X, and by compactness there is then a finite subcovering

$$X = \bigcup_{k=1}^{m} B_{\varepsilon_{x_k}}(x_k).$$

We now show that

$$\delta = \min\{\varepsilon_{x_1}, \ldots, \varepsilon_{k_m}\},$$

which is clearly positive, satisfies the requirement of the lemma. Suppose then that $S \subseteq X$ is a subset with $\operatorname{diam}(S) < \delta$, and we may assume $S \neq \emptyset$, so let us fix an element $y \in S$. Since $y \in X$ and X is covered by $\{B_{\varepsilon_k}(x)\}_{1\leq k\leq n}$ we may find an $x = x_{k_0}$, with corresponding $\varepsilon = \varepsilon_{k_0}$, such that $y \in B_\varepsilon(x)$. It now follows, for any $y' \in S$, that $d(y, y') \leq \operatorname{diam}(S) < \delta$, and thus $y' \in B_\delta(y)$. However,

$$B_\delta(y) \subseteq B_\varepsilon(y) \subseteq B_{2\varepsilon}(x) \subseteq U_x$$

(as the reader is invited to verify) and thus $y' \in U_x$. As $y' \in S$ was arbitrary it follows that $S \subseteq U_x$, as required. □

A δ as in the statement above is called a *Lebesgue number* for the given covering.

Theorem 4.7 (Heine-Cantor) *A continuous function $f : X \to Y$ from a compact metric space X to an arbitrary metric space Y is uniformly continuous.*

Proof Given $\varepsilon > 0$ consider, for each $y \in Y$, the set

$$U_y = f^{-1}(B_{\frac{\varepsilon}{2}}(y))$$

which is open since f is continuous. Clearly, $x \in U_{f(x)}$ for all $x \in X$, and thus $\{U_y\}_{y\in X}$ is an open covering of X. We may now choose a Lebesgue number $\delta > 0$ for that covering. Suppose that

$$d(x, x') < \delta$$

for points $x, x' \in X$. Then, $\operatorname{diam}(\{x, x'\}) < \delta$ and thus $\{x, x'\} \subseteq f^{-1}(B_{\varepsilon/2}(y))$ for some $y \in Y$. That means that $d(f(x), y) < \varepsilon/2$ and that $d(f(x'), y) < \varepsilon/2$.

Applying the triangle inequality yields

$$d(f(x), f(x')) \leq d(f(x), y) + d(y, f(x')) < \varepsilon$$

as needed. □

Since uniform continuity always entails continuity, the result above implies that for a function $f : X \to Y$ between metric spaces with a compact domain the concepts of continuity and of uniform continuity coincide.

Exercises

Exercise 4.21 Prove that the identity function $\mathrm{id} : X \to X$ on any metric space is uniformly continuous, and prove that the composition of two uniformly continuous functions is uniformly continuous. Repeat the exercise for non-expanding mappings instead of uniformly continuous ones.

Exercise 4.22 For a function $f : X \to Y$ between metric spaces show that if f is non-expanding, then f is uniformly continuous. Show as well that if f is uniformly continuous, then f is continuous.

Exercise 4.23 Construct an example of a continuous function $f : X \to Y$ between metric spaces that is not uniformly continuous, and an example where f is uniformly continuous, but not non-expanding.

Exercise 4.24 Let $f : \mathbb{R} \to \mathbb{R}$ be a differentiable function with $|f'(x)| \leq 1$ for all $x \in \mathbb{R}$. Prove that f is non-expanding, when considering the Euclidean metric.

Exercise 4.25 Let X be a metric space and $\varepsilon > 0$. Show that if $d(x, y) < \varepsilon$, then $B_\varepsilon(y) \subseteq B_{2\cdot\varepsilon}(x)$.

Exercise 4.26 Prove that if $f : X \to Y$ is non-expanding, then $\mathrm{diam}(f(S)) \leq \mathrm{diam}(S)$ for all $S \subseteq X$. Does the converse hold?

Exercise 4.27 Prove that if $f : X \to Y$ is uniformly continuous, then for every $\varepsilon > 0$ there exists a $\delta > 0$ such that $\mathrm{diam}(f(S)) < \varepsilon$ for all $S \subseteq X$ with $\mathrm{diam}(S) < \delta$. Does the converse hold?

Exercise 4.28 Prove that for all subsets S_1, S_2 of a metric space X, if $S_1 \subseteq S_2$, then $\mathrm{diam}(S_1) \leq \mathrm{diam}(S_2)$. Does the converse hold?

Exercise 4.29 Prove that $\mathrm{diam}(B_\varepsilon(x)) \leq 2\varepsilon$ for all $\varepsilon > 0$ and for all points x in a metric space X.

Exercise 4.30 Give an example of a metric space X, a point $x \in X$, and $\varepsilon > 0$ such that the equality $\mathrm{diam}(B_\varepsilon(x)) = 2\varepsilon$ fails.

4.4 Complete Metric Spaces

This section is devoted to completeness in a metric space. A metric space X is said to be complete when every Cauchy sequence in it converges. Thus, in a sense, X is complete when every sequence in X that should converge actually does converge. The section starts with a discussion of the concept of Cauchy sequences in general metric spaces and then considers the completeness of $\mathbb{R}$. The reader may already be familiar with the completeness of the real numbers, a property that is commonly stated in the form of the least upper bound principle: every non-empty bounded above set of real numbers has a supremum (for more details see Sect. 1.2.19 of the Preliminaries). We show below that the least upper bound property implies the metric completeness of $\mathbb{R}$, namely that any Cauchy sequence in $\mathbb{R}$ converges (when $\mathbb{R}$ is given the Euclidean metric). We then examine the metric spaces introduced above, together with an investigation of their completeness. Next we establish two fundamental theorems of complete metric spaces, namely the Banach Fixed-Point Theorem and Baire's Theorem. The importance of these results is difficult to over emphasize, particularly in the context of this book. As will be shown in Chap. 5, the Banach Fixed-Point Theorem can be used to iteratively solve certain differential equations. Some of the consequences of Baire's Theorem in the theory of Banach spaces will also be given. This chapter then closes with a description of a completion process, i.e., a way to turn any metric space, complete or not, into a complete one while altering it as little as possible. The construction we present is that of Cantor's using Cauchy sequences.

4.4.1 Complete Metric Spaces

Given a sequence $\{x_m\}_{m\geq 1}$ in a metric space X, the statement

$$\exists x_0 \in X \, \forall \varepsilon > 0 \, \exists N \in \mathbb{N} \, \forall m \geq N : \; d(x_m, x_0) < \varepsilon$$

is nothing but the claim that $x_m \to x_0$. By interchanging the first two quantifiers in the statement above we arrive at the statement

$$\forall \varepsilon > 0 \, \exists x_\varepsilon \in X, \; N \in \mathbb{N} \, \forall m \geq N : \; d(x_m, x_\varepsilon) < \varepsilon$$

which we call the *global Cauchy condition*. It expresses something considerably weaker than the existence of a limit. Namely, to establish convergence one needs to manufacture an appropriate $N \in \mathbb{N}$ for any given $\varepsilon > 0$ *without* altering the limiting point. With the global Cauchy condition on the other hand one is first given an $\varepsilon > 0$, and may then manufacture the limit point x_ε, and a corresponding $N \in \mathbb{N}$. Thus the global Cauchy condition expresses the existence of a sort of varying limit of the sequence. In particular, if the sequence converges to x_0, then the global Cauchy condition is satisfied simply by taking $x_\varepsilon = x_0$.

A simple analysis of the situation reveals that the x_ε can be entirely removed from the global Cauchy condition, as follows. Suppose that the global Cauchy condition is satisfied and let $\varepsilon > 0$ be given. There exist then an $x_\varepsilon \in X$ and an $N \in \mathbb{N}$ so that $d(x_m, x_\varepsilon) < \varepsilon/2$, for all $m \geq N$. It then follows that

$$d(x_k, x_m) \leq d(x_k, x_\varepsilon) + d(x_\varepsilon, x_m) \leq \frac{\varepsilon}{2} + \frac{\varepsilon}{2} = \varepsilon$$

for all $k, m \geq N$. Conversely, suppose that for all $\varepsilon > 0$ there exists an $N \in \mathbb{N}$ such that

$$d(x_k, x_m) < \varepsilon$$

for all $k, m \geq N$. Then, taking $x_\varepsilon = x_N$, it follows that

$$d(x_m, x_\varepsilon) = d(x_m, x_N) < \varepsilon$$

for all $m \geq N$, and so the global Cauchy condition is satisfied. This discussion is summarized as follows.

Definition 4.11 A sequence $\{x_m\}_{m\geq 1}$ in a metric space X is a *Cauchy sequence* if it satisfies the *Cauchy condition*:

$$\forall \varepsilon > 0 \, \exists N \in \mathbb{N} \, \forall k, m \geq N : \, d(x_k, x_m) < \varepsilon.$$

Equivalently, a sequence $\{x_m\}_{m\geq 1}$ is a Cauchy sequence if it satisfies the *global Cauchy condition*

$$\forall \varepsilon > 0 \, \exists x_\varepsilon \in X, \; N \in \mathbb{N} \, \forall m \geq N : \, d(x_m, x_\varepsilon) < \varepsilon.$$

The familiar fact that every convergent sequence in $\mathbb{R}$ (when endowed with the Euclidean metric) is a Cauchy sequence is now nothing but the remark, made above, that the global Cauchy condition is trivially implied by the existence of a limit. The converse in $\mathbb{R}$ is well-known to be true (and we prove that fact below) but for general metric spaces the converse may fail. To see that, start with any convergent sequence $x_m \to x_0$ in a metric space X satisfying $x_m \neq x_0$, for all $m \geq 1$, and consider the same sequence $\{x_m\}_{m\geq 1}$ in the subspace $X - \{x_0\}$. Clearly, $\{x_m\}_{m\geq 1}$ satisfies the Cauchy condition (indeed, it is a convergent sequence in X). However, it only converges in X since if it converged in $X - \{x_0\}$ too, say to the point $y \in X - \{x_0\}$, then it would also converge to y in X, contradicting uniqueness of limits in a metric space.

Definition 4.12 A metric space X in which every Cauchy sequence admits a limit *in* X is said to be a *complete metric space*.

As promised above we now closely examine the completeness of the reals, which, as we will see, is a pivotal result.

Example 4.12 (*The completeness of* $\mathbb{R}$) The space $\mathbb{R}$ with the Euclidean metric $d(x, y) = |x - y|$ is a most fundamental example of a complete metric space. Recall (e.g., Sect. 1.2.19 of the Preliminaries) that the real numbers $\mathbb{R}$ are defined to be any Dedekind complete ordered field.

Given a Cauchy sequence $\{x_m\}_{m\geq 1}$ of real numbers, we first note that $\{x_m\}_{m\geq 1}$ is bounded by some number M, i.e., $|x_m| < M$ for all $m \in \mathbb{N}$. Indeed, for $\varepsilon = 1$ there exists an $N \in \mathbb{N}$ such that $|x_k - x_m| < 1$ for all $k, m \geq N$, and thus we found a bounded tail of the sequence, and thus the entire sequence is bounded. Let $S \subseteq \mathbb{R}$ be the subset consisting of all $x \in \mathbb{R}$ such that the inequality $x < x_m$ holds for almost all $m \in \mathbb{N}$ (meaning that it holds for all $m \in \mathbb{N}$ with at most finitely many exceptions). Clearly, for our chosen upper bound M we have that $-M \in S$ and $M \notin S$, and thus S is non-empty and bounded above. By the least upper bound principle in $\mathbb{R}$ the set S admits a supremum L. We will proceed to show that L is in fact the limit of the given sequence. Given $\varepsilon > 0$ let $N \in \mathbb{N}$ with $|x_k - x_m| < \varepsilon/2$ for all $k, m \geq N$. In particular, taking $m = N$, we see that

$$x_m - \frac{\varepsilon}{2} < x_k < x_m + \frac{\varepsilon}{2}$$

holds for all $k \geq N$. These inequalities imply, by definition of S, that

$$x_m - \frac{\varepsilon}{2} \in S, \quad \text{while} \quad x_m + \frac{\varepsilon}{2} \notin S.$$

Since L is the least upper bound of S it now follows that

$$x_m - \frac{\varepsilon}{2} \leq L \leq x_m + \frac{\varepsilon}{2}.$$

Combining all of the inequalities above, it follows that

$$|x_k - L| \leq |x_k - x_m| + |x_m - L| \leq \frac{\varepsilon}{2} + \frac{\varepsilon}{2} = \varepsilon$$

for all $k > N$, proving that $x_m \to L$. $\mathbb{R}$ is thus metrically complete.

Remark 4.4 The amount of work put into this example is quite adequately justified by noticing that the completeness of $\mathbb{R}$ is used in establishing the completeness of all of the forthcoming examples of complete metric spaces.

Remark 4.5 It should be noted that Dedekind completeness is a stronger property than metric completeness, in the following sense. While every Dedekind complete ordered field is metrically complete, there exist metrically complete ordered fields which are not Dedekind complete. None of these examples would be Archimedean though. These issues however are not particularly relevant to us here.

Example 4.13 In contrast to the situation with the real numbers, the metric space $\mathbb{Q}$ of rational numbers with $d(x, y) = |x - y|$ is not complete. Indeed, by the density of $\mathbb{Q}$ in $\mathbb{R}$, for every irrational α there is a sequence of rational numbers with $x_m \to \alpha$ in $\mathbb{R}$. Such a sequence is thus Cauchy in both $\mathbb{R}$ and $\mathbb{Q}$ but fails to converge in $\mathbb{Q}$.

Before examining the completeness property for the metric spaces introduced above, we remark that completeness is a metric invariant. In fact, we establish a slightly stronger result.

Proposition 4.2 (Uniformly Continuous Functions Preserve Cauchy Sequences) *If $f : X \to Y$ is a uniformly continuous function between metric spaces and $\{x_m\}_{m\geq 1}$ is a Cauchy sequence in X, then $\{f(x_m)\}_{m\geq 1}$ is a Cauchy sequence in Y.*

Proof Given $\varepsilon > 0$ let $\delta > 0$ be such that

$$d(x, x') < \delta \implies d(f(x), f(x')) < \varepsilon.$$

Since $\{x_m\}_{m\geq 1}$ is Cauchy, there exists an $N \in \mathbb{N}$ such that

$$k, m \geq N \implies d(x_k, x_m) < \delta.$$

Combining the two implications yields

$$k, m \geq N \implies d(f(x_k), f(x_m)) < \varepsilon,$$

as needed for showing that $\{f(x_m)\}_{m\geq 1}$ is Cauchy. □

Corollary 4.5 *Let $f : X \to Y$ be an invertible continuous function between metric spaces such that $f^{-1} : Y \to X$ is uniformly continuous. If X is complete, then so is Y.*

Proof Let $\{y_m\}_{m\geq 1}$ be a Cauchy sequence in Y. Then $\{f^{-1}(y_m)\}_{m\geq 1}$ is a Cauchy sequence in X and since X is complete the sequence converges: $f^{-1}(y_m) \to x_0$. Since a continuous function preserves limits of sequences, we may now apply f to conclude that $y_m \to f(x_0)$. □

Corollary 4.6 *Completeness is a metric invariant.*

Example 4.14 (Completeness of $\mathbb{R}^n$) For the metric space $\mathbb{R}^n$ with the Euclidean metric

$$d(x, y) = \left(\sum_{k=1}^{n} (x_k - y_k)^2\right)^{1/2}$$

we already saw that $x^{(m)} \to x$ if, and only if, $x_k^{(m)} \to x_k$ for each $1 \leq k \leq n$. To show that $\mathbb{R}^n$ is complete, suppose that $\{x^{(m)}\}_{m\geq 1}$ is a Cauchy sequence. Then, for each

$1 \leq k \leq n$, the sequence $\{x_k^{(m)}\}_{m\geq 1}$ is also Cauchy (since $|x_k^{(m)} - x_k^{(m')}| \leq d(x, y)$) and since $\mathbb{R}$ is complete, we conclude that $x_k^{(m)} \to x_k$ for all $1 \leq k \leq n$, and therefore that $x^{(m)} \to (x_1, \ldots, x_n)$. Thus, every Cauchy sequence in $\mathbb{R}^n$ converges, as claimed.

Example 4.15 (Completeness of $\mathbb{C}^n$*)* The metric space $\mathbb{C}^n$ with distance given by

$$d(x, y) = \left(\sum_{k=1}^{n} |x_k - y_k|^2 \right)^{1/2}$$

is complete. It was already remarked that $\mathbb{C}^n$ is isometric to $\mathbb{R}^{2n}$, and thus, since completeness is a metric invariant, the completeness of $\mathbb{R}^{2n}$ implies that of $\mathbb{C}^n$.

Example 4.16 For a closed interval $I = [a, b]$ the space $C(I, \mathbb{R})$ of continuous functions on I, with the metric structure induced by the L_∞ norm, is complete. To see that, recall that convergence with respect to this metric is uniform convergence of sequences of functions. If $\{x_m\}_{m\geq 1}$ is a Cauchy sequence of continuous real-valued functions on I, then one easily sees that for each $t \in [a, b]$ the sequence $\{x_m(t)\}_{m\geq 1}$ is a Cauchy sequence in $\mathbb{R}$ with respect to the Euclidean metric. As $\mathbb{R}$ is complete it follows that $x_m(t) \to x(t)$, showing that $\{x_m\}_{m\geq 1}$ converges point-wise to a function $x : [a, b] \to \mathbb{R}$, and in fact the convergence is easily seen to be uniform. To conclude we recall the well-known fact that if a sequence of continuous functions converges uniformly to a function x, then x is continuous. Thus, $x \in C(I, \mathbb{R})$ and is the uniform limit of the given sequence.

Example 4.17 In contrast to the L_∞ case, for all $1 \leq p < \infty$ the pre-L_p space $C([a, b], \mathbb{R})$ is not complete. For simplicity let us consider the case $[a, b] = [0, 1]$. Let

$$x_m(t) = \begin{cases} 1 & \text{if } 0 \leq x \leq \frac{1}{2}, \\ mx - \frac{1}{2m} - 1 & \text{if } \frac{1}{2} \leq x \leq \frac{1}{2} + \frac{1}{m}, \\ 0 & \text{if } \frac{1}{2} + \frac{1}{m} \leq x \leq 1. \end{cases}$$

Then, for $k \geq m$

$$\|x_m - x_k\|_p^p = \int_{\frac{1}{2}}^{\frac{1}{2}+\frac{1}{n}} dt\, |x_m(t) - f_k(t)|^p \leq \frac{1}{n}$$

and thus $\{x_m\}_{m\geq 1}$ is Cauchy. However, if this sequence converged to a continuous function $x : [0, 1] \to \mathbb{R}$, then

$$\int_0^{\frac{1}{2}} dt\ |x(t) - x_m(t)|^p \leq \|x - x_m\|_p^p \to 0$$

and thus $x(t) = 1$ for all $0 \leq t < 1/2$. Similarly $x(t) = 0$ for all $1/2 < t \leq 1$, but then x can not be continuous.

Example 4.18 Going back to $C(I, \mathbb{R})$, $I = [a, b]$ and $a < b$, with the metric obtained from the L_∞ norm (for which convergence is equivalent to uniform convergence), which was shown to be complete, consider the subspace $P \subseteq C(I, \mathbb{R})$ consisting only of the polynomial functions. As a subspace of the complete space $C(I, \mathbb{R})$ the space P is not complete. To see that, recall that the polynomials

$$p_m(t) = \sum_{k=0}^{m} \frac{t^k}{k!}$$

converge uniformly to the exponential function e^t, which is not a polynomial. Thus the sequence $\{p_m\}_{m \geq 1}$ is Cauchy (since it converges in $C(I, \mathbb{R})$) but it does not converge to a polynomial and thus does not converge in P at all (since if it would, then it would have two different limits in $C(I, \mathbb{R})$).

Example 4.19 The space ℓ_p (of all absolutely p-power summable sequences of, say, real numbers) is complete. To see that, let $\{x^{(m)}\}_{m \geq 1}$ be a Cauchy sequence in ℓ_p. Since $|x_k^{(m)} - x_k^{(n)}| \leq \|x^{(m)} - x^{(n)}\|_p$, for all $k \geq 1$, it follows readily that $\{x_k^{(m)}\}_{m \geq 1}$ is a Cauchy sequence in $\mathbb{R}$, and thus it converges to a point $x_k \in \mathbb{R}$. Thus $\{x^{(m)}\}_{m \geq 1}$ converges point-wise to $x = (x_1, \ldots, x_k, \ldots)$. The completeness of ℓ_p will follow by showing that the convergence is also in the ℓ_p norm. Firstly, to verify that $x \in \ell_p$, note that since $\{x^{(m)}\}_{m \geq 1}$ is Cauchy, it is bounded, say by M. Now, for all $K \geq 1$ and $m \geq 1$

$$\left(\sum_{k=1}^{K} (x_k^{(m)})^p \right)^{1/p} \leq \|x^{(m)}\|_p \leq M.$$

Allowing m to tend to ∞, and then for K to tend to ∞ shows that $\|x\|_p \leq M$, and thus $x \in \ell_p$. Finally, to show that $x_m \to x$ in the ℓ_p norm, let $\varepsilon > 0$ be given. Let $N \in \mathbb{N}$ with

$$\|x^{(m)} - x^{(n)}\|_p < \varepsilon$$

for all $m, n \geq N$. We now have, for all $K \geq 1$ and $m, n \geq N$, that

$$\left(\sum_{k=1}^{K}(x_k^{(m)} - x_k^{(n)})^p\right)^{1/p} \leq \|x^{(m)} - x^{(n)}\|_p < \varepsilon$$

and thus, letting n tend to infinity, and then K tend to infinity, it follows that

$$\|x^{(m)} - x\|_p < \varepsilon$$

for all $m > N$, as required.

4.4.2 Banach's Fixed-Point Theorem

A fixed-point for a function $f : X \to X$ is a point $x_0 \in X$ such that $f(x_0) = x_0$. The importance of fixed-points lies in the ability to phrase the solutions of many important problems as fixed-points for suitable functions (this technique will be demonstrated in Chap. 5).

The following metric condition will be shown to guarantee the existence of a unique fixed-point for a function f as above, provided X is a non-empty complete metric space.

Definition 4.13 A function $f : X \to Y$ between metric spaces is called a *contraction* if there exists a real number α with $0 \leq \alpha < 1$ such that

$$d(f(x_1), f(x_2)) \leq \alpha d(x_1, x_2)$$

for all $x_1, x_2 \in X$.

We note at once that a function need not be a contraction in order to posses fixed-points. An extreme example is the identity function $\text{id} : X \to X$ on any metric space. It is not a contraction yet every point is a fixed-point for it.

Theorem 4.8 (Banach Fixed-Point Theorem) *If X is a complete non-empty metric space and $f : X \to X$ is a contraction, then f has a unique fixed-point $x_0 \in X$.*

Proof To prove the existence of a fixed-point, let x_1 be an arbitrary element in X and consider the sequence $x_2 = f(x_1)$, $x_3 = f(x_2)$, and in general $x_{m+1} = f(x_m)$. If this sequence converges to some limit $x_0 \in X$, then, since continuous functions preserve limits of sequences, we will have that

$$f(x_0) = f(\lim_{m\to\infty} x_m) = \lim_{m\to\infty} f(x_m) = \lim_{m\to\infty} x_{m+1} = \lim_{m\to\infty} x_m = x_0$$

and so x_0 is a fixed-point of f. To show that the sequence does indeed converge, we employ the completeness of X and proceed to show that the sequence is Cauchy. Since the function f is a contraction, we know that there exists an $0 \leq \alpha < 1$ such that

$$d(f(x), f(y)) \leq \alpha d(x, y)$$

for all $x, y \in X$. We need to estimate the quantities $d(x_k, x_m)$ for the sequence defined above, and we notice first that for all $x, y \in X$, the triangle inequality yields

$$\begin{aligned} d(x, y) &\leq d(x, f(x)) + d(f(x), f(y)) + d(f(y), y) \\ &\leq d(x, f(x)) + \alpha d(x, y) + d(f(y), y) \end{aligned}$$

and thus

$$d(x, y) \leq \frac{d(x, f(x)) + d(y, f(y))}{1 - \alpha}$$

which immediately shows that if x and y are both fixed-points of f, then $d(x, y) = 0$, and thus $x = y$, so the uniqueness clause is proven.

A straightforward application of the latter inequality shows that

$$d(x_k, x_m) \leq \frac{\alpha^k + \alpha^m}{1 - \alpha} d(x_1, f(x_1))$$

immediately establishing that $\{x_m\}_{m \geq 1}$ is a Cauchy sequence. □

4.4.3 Baire's Theorem

The result we present below has numerous important implications in analysis in general and in functional analysis and Banach space theory in particular. Baire's theorem can be interpreted as stating that a non-empty complete metric space can not be *small* when size is topologically measured as follows. The set $\mathbb{Q}$ of rational numbers is countable while $\mathbb{R}$ is uncountable and thus, in the sense of cardinality of sets, $\mathbb{Q}$ is negligible compared to $\mathbb{R}$. However, from a topological point-of-view, since the closure of $\mathbb{Q}$ is all of $\mathbb{R}$, the set $\mathbb{Q}$ is quite large. The opposite situation is that of a *nowhere dense* set, namely a set S whose closure has empty interior, and is thus a (topologically speaking) very small set. Sets which are countable unions of nowhere dense sets are called *meager* sets. The claim of Baire's Theorem is that a non-empty complete metric space is never meager.

Remark 4.6 Many of the definitions below can be stated in the greater generality of topological spaces. However, we keep the discussion rooted in metric spaces since the added generality is not of importance here.

Definition 4.14 A subset $S \subseteq X$ in a metric space X is said to be *nowhere dense* if the interior of its closure is empty. Equivalently, S is nowhere dense when $\overline{S}$ contains no non-empty open subsets.

Nowhere dense sets are, in a sense, very small. For instance, if x is not an isolated point, i.e., $\{x\}$ is not open, then $\{x\}$ is nowhere dense. It is obvious that a subset of a nowhere dense subset is itself nowhere dense, and that a finite union of nowhere dense subsets is itself nowhere dense. However, a countable union of nowhere dense subsets need not be nowhere dense. For instance, in $\mathbb{R}$ with the Euclidean metric no point is isolated and thus every singleton $\{x\}$ is nowhere dense, and consequently every finite subset of $\mathbb{R}$ is nowhere dense. However, $\mathbb{Q}$ (which is a countable union of singleton sets, and thus of nowhere dense sets) is not nowhere dense. In fact, it is everywhere dense, i.e., its closure is the entire space $\mathbb{R}$.

Definition 4.15 A subset $S \subseteq X$ in a metric space X is said to be *meager* if it is a countable union of nowhere dense subsets.

Clearly, every nowhere dense subset is meager, but not vice-versa (e.g., $\mathbb{Q}$ is meager in $\mathbb{R}$ but not nowhere dense). It is again left for the reader to establish that a subset of a meager set is itself meager, and that a countable union of meager subsets is meager.

Since X is a subset of itself we may speak of X itself being meager. We may now state the main theorem.

Theorem 4.9 (Baire) *A complete non-empty metric space is not meager.*

Proof Suppose that X is a complete, non-empty, and meager metric space. Then

$$X = \bigcup_{m=1}^{\infty} S_m$$

where each $S_m \subseteq X$ is nowhere dense. By noting that the closure of a nowhere dense subset is itself nowhere dense, we may further assume (by taking closures) that each S_m is closed. Since X is non-empty X is not nowhere dense, and thus, for all $m \geq 1$, one has that $S_m \neq X$.

It follows that $X - S_1$ is a non-empty open subset of X, and so there exists a point $x_1 \in X$ and $\varepsilon_1 > 0$ with

$$B_{\varepsilon_1}(x_1) \subseteq X - S_1.$$

Since S_2 does not contain any open subset, it certainly does not contain $B_{\varepsilon_1}(x)$. Thus the set $B_{\varepsilon_1}(x_1) \cap (X - S_2)$, which is open, is not empty. Therefore, there exists a point $x_2 \in X$ and $\varepsilon_2 > 0$ with

$$B_{\varepsilon_2}(x_2) \subseteq B_{\varepsilon_1}(x_1) \cap (X - S_2).$$

Continuing in this manner we obtain, for each $m \geq 1$, a point $x_m \in X$ and $\varepsilon_m > 0$ such that

$$B_{\varepsilon_{m+1}}(x_{m+1}) \subseteq B_{\varepsilon_m}(x_m) \cap (X - S_m).$$

Notice that we may assume that $\varepsilon_m < 1/m$ holds for all $m \geq 1$, which readily implies that $\{x_m\}_{m\geq 1}$ is a Cauchy sequence, which thus converges to a limit point $x_0 \in X$. Our next claim is that the limit point x_0 belongs to $B_{\varepsilon_m}(x_m)$ for each $m \geq 1$. Indeed, by the assumption that $\varepsilon_m \leq 1/m$, it follows, for all $k \geq m$, that $B_{\varepsilon_k}(x_k)$ is contained in the closure of $B_{\varepsilon_{m+1}}(x_{m+1})$, which in turn is contained in $B_{\varepsilon_m}(x_m)$. Thus, a tail of the sequence is contained in the closure of $B_{\varepsilon_{m+1}}(x_{m+1})$, which is closed, and thus the limit is contained in it too, and thus the limit is in $B_{\varepsilon_m}(x_m)$. Now, since $\{S_m\}_{m\geq 1}$ covers X, we must have $x_0 \in S_{m_0}$ for some $m_0 \geq 1$. But then, $x_0 \notin X - S_{m_0}$ and thus $x_0 \notin B_{\varepsilon_0+1}(x_{m+1})$, a contradiction. □

Corollary 4.7 *A complete non-empty metric space X with no isolated points must be uncountable.*

Proof If X were countable, then since

$$X = \bigcup_{x \in X} \{x\}$$

and each $\{x\}$ is nowhere dense it would follow that X is meager. □

Corollary 4.8 *The set $\mathbb{R}$ of real numbers is uncountable.*

Proof When endowed with the Euclidean metric, the space $\mathbb{R}$ is complete and has no isolated points. □

4.4.4 Completion of a Metric Space

If X is a metric space which is not complete, then it is because there exist Cauchy sequences in it that fail to converge in X. Quite often though the space X is seen to be embedded inside a larger metric space which is complete. This is the situation for instance for $\mathbb{Q}$ as a metric subspace of $\mathbb{R}$ with the Euclidean metric. There are however other situation where it is not a-priori clear that the metric space in question embeds in a complete one. For instance, the space $C(I, \mathbb{R})$ with the metric obtained from the L_2 norm is not complete (as we have seen) and it is not evident that it is embedded in a larger, complete, space.

It is thus natural to ask if any metric space can be completed. That is, whether every metric space embeds in a potentially (much) larger complete metric space. Of course, one would have to set some expectations from such a completion, primarily that it is not too wasteful. In more detail, we are willing to accept that the completion may need to be larger than the original space, but we do not want it to be any larger than it must be. The following definition turns this rather vague formulation into a precise condition.

Definition 4.16 Let X be a metric space. A complete metric space Y together with an isometry $\iota : X \to Y$ is said to be a *metric completion* (or simply a *completion*) of

X if $\iota(X)$ is dense in Y. It is common to speak of Y as a completion of X, leaving $\iota : X \to Y$ implicit, and it is also common to write $\hat{X}$ for a completion.

Example 4.20 Clearly, the inclusion $\iota : \mathbb{Q} \to \mathbb{R}$ expresses $\mathbb{R}$ as a completion of $\mathbb{Q}$, when both sets are endowed with the metric $d(x, y) = |x - y|$.

The following theorem states that uniformly continuous functions with complete codomains can be extended uniquely to a completion of the domain. This is a very important property of metric completions.

Theorem 4.10 *Let $f : X \to Y$ be a uniformly continuous function between metric spaces, with Y complete. If $\iota : X \to \hat{X}$ is a completion, then there exists a unique uniformly continuous function $F : \hat{X} \to Y$ which extends f, i.e., such that $f = F \circ \iota$.*

Proof We present here only the outline of the proof, inviting the reader to provide further details.

1. To construct an extension $F : \hat{X} \to Y$ let $x_0 \in \hat{X}$ be arbitrary and consider any sequence $\{x_m\}_{m\geq 1}$ in X with $\iota(x_m) \to x_0$. This can be done since $\iota(X)$ is dense in $\hat{X}$.
2. Since $\{\iota(x_m)\}_{m\geq 1}$ converges, it is a Cauchy sequence, and since ι is an isometry, $\{x_m\}_{m\geq 1}$ is Cauchy as well.
3. Since a uniformly continuous function preserves Cauchy sequences, it follows that the sequence $\{f(x_m)\}_{m\geq 1}$ is Cauchy in Y.
4. Since Y is complete, every Cauchy sequence converges, and thus $f(x_m) \to y_0$, for some $y_0 \in Y$, and we denote $F(x_0) = y_0$.
5. At this point one needs to verify that $F(x_0)$ is independent of the chosen sequence $\{x_m\}_{m\geq 1}$. This can be done by showing that for any other sequence $\{x'_m\}_{m\geq 1}$ in X with $\iota(x'_m) \to x_0$,
$$d(\lim_{k\to\infty} f(x_m), \lim_{k\to\infty} f(x'_m)) = 0$$
(by appealing to the continuity of the distance function).
6. The verification that F extends f, namely that $F(\iota(x)) = f(x)$ for all $x \in X$, is immediate by considering the constant sequence $\{x_m\}_{m\geq 1}$, $x_m = x$.
7. One then needs to verify that F is uniformly continuous. Consequently, F is continuous.
8. Finally, the uniqueness of F follows since any two continuous functions that agree on a dense subset, agree everywhere. □

Remark 4.7 The ability to extend a uniformly continuous function to a completion of the domain does not hold for arbitrary continuous functions. For instance, take the set $S = \{1/n \mid n \in \mathbb{N}\}$ as a metric subspace of $\mathbb{R}$ under the Euclidean metric. Topologically, it is a discrete space and thus any function $f : S \to Y$, to any metric space Y, is continuous. Such a function merely corresponds to a sequence in Y. Now, let $\hat{S} = S \cup \{0\}$ and notice that with the inclusion function $\iota : S \to \hat{S}$, we obtain a completion of S. The space $\hat{S}$ is not discrete and indeed one can easily verify that a continuous function $F : \hat{S} \to Y$ corresponds to a sequence in Y together with its limit point. In other words, while continuous functions $f : S \to Y$ correspond to arbitrary sequences in Y, continuous functions $F : \hat{S} \to Y$ correspond to convergent sequences in Y. Clearly, there are plenty of example of complete metric spaces Y where not every sequence converges, and thus not every continuous function $f : S \to Y$ extends to a continuous function $F : \hat{S} \to Y$.

We now pose and answer two natural questions about completions of a metric space X, namely does a completion always exist and whether it is unique. We first address the question of uniqueness. Suppose that $\iota : X \to Y$ is a completion. Given any bijection $g : Y \to Y'$, with Y' an arbitrary set, the metric structure of Y can be transferred to a metric structure on Y' by defining $d : Y' \times Y' \to \mathbb{R}_+$ by $d(y_1', y_2') = d(g^{-1}(y_1'), g^{-1}(y_2'))$, and then g is a global isometry. Informally, there is no essential difference between Y and Y', and thus it is expected that Y' can also serve as a completion for X. This is indeed the case, as the reader can check that the composition $g \circ \iota : X \to Y'$ is a completion. It thus follows that a completion is never unique (except when it is empty), but, more importantly, we are guided to ask a more refined question: Given two completions $\iota : X \to Y$ and $\iota' : X \to Y'$ for the same metric space X, are Y and Y' isometric? The affirmative answer is embodied in the following result.

Theorem 4.11 *Let X be a metric space. If $\iota : X \to Y$ and $\iota' : X \to Y'$ are completions of X, then there exists a global isometry $\rho : Y \to Y'$.*

Proof Since $\iota' : X \to Y'$ is an isometry, it is in particular uniformly continuous, and thus ι' extends uniquely to a uniformly continuous function $\rho : Y \to Y'$, satisfying that $\rho \circ \iota = \iota'$. Reversing the roles of ι and ι', the same argument yields a uniformly continuous function $\rho' : Y' \to Y$ with $\rho' \circ \iota' = \iota$. Let us now consider $\rho' \circ \rho : Y \to Y$, for which we have that $\rho' \circ \rho \circ \iota = \rho' \circ \iota' = \iota$. Thus, $\rho' \circ \rho : Y \to Y$ is a uniformly continuous function which extends ι. However, quite trivially, the identity function $\mathrm{id}_Y : Y \to Y$ satisfies $\mathrm{id}_Y \circ \iota = \iota$, namely it too is a uniformly continuous function extending ι. But, such an extension was proven above to be unique, and thus we must conclude that $\rho' \circ \rho = \mathrm{id}_Y$. A similar argument reveals that $\rho \circ \rho' = \mathrm{id}_{Y'}$, and thus that ρ is bijective and ρ' is its inverse. Finally, due to the continuity of the distance function, for all $y_1, y_2 \in Y$, choosing sequences $\{a_m\}_{m\geq 1}$ and $\{b_m\}_{m\geq 1}$ with $\iota(a_m) \to y_1$ and $\iota(b_m) \to y_2$,

$$d_Y(y_1, y_2) = \lim_{m\to\infty} d_Y(\iota(a_m), \iota(b_m)) = \lim_{m\to\infty} d_X(a_m, b_m)$$

while

$$\begin{aligned} d_{Y'}(\rho(y_1), \rho(y_2)) &= \lim_{m\to\infty} d_{Y'}((\rho \circ \iota)(a_m), (\rho \circ \iota)(b_m)) \\ &= \lim_{m\to\infty} d_{Y'}(\iota'(a_m), \iota'(b_m)) = \lim_{m\to\infty} d_X(a_m, b_m). \end{aligned}$$

In short then, $d_Y(y_1, y_2) = d_{Y'}(\rho(y_1), \rho(y_2))$, namely ρ is an isometry, and, as seen above, a global one. □

The result above shows that all completions of a metric space X are isometric and thus that the completion is *unique up to an isomorphism.* In light of the preceding discussion, we can not hope for anything more than that.

We now turn to the existence of a completion of an arbitrary metric space X. Since a completion is only unique up to an isometry, and thus if one (non-empty) completion of X exists, then infinitely many different completions of X exist, it should not be surprising that there are in the literature different constructions of embeddings $\iota : X \to \hat{X}$, yielding different (yet isometric) completions. The one we present is due to Cantor. The reader may wish to refer to the construction of the reals, outlined in Sect. 1.1, since it is closely related to the general completion construction below, only framed in a more familiar and less abstract setting.

Let us fix a metric space X and construct a completion for it in a series of steps, leaving it to the reader to provide the proofs. If X is not complete, then there exists a Cauchy sequence $\{x_m\}_{m\geq 1}$ which fails to have a limit in X. Intuitively, completing X entails adjoining to X formal limits for all such Cauchy sequences. If we denote the formal limit of $\{x_m\}_{m\geq 1}$ by $[\{x_m\}]$, then we are led to define $\hat{X}$ as the union

$$X \cup \{[\{x_m\}] \mid \{x_m\}_{m\geq 1} \text{ is a Cauchy sequence in } X \text{ which fails to converge}\}.$$

However, two different Cauchy sequences $\{x_m\}_{m\geq 1}$ and $\{y_m\}_{m\geq 1}$ may converge to the same point (for instance, there are many different sequences of rationals that converge to, e.g., $\sqrt{2}$), and thus an equivalence relation needs to be introduced. There is also a simplifying observation, namely that we may identify an element $x \in X$ with the constant sequence $\{x_m = x\}_{m\geq 1}$ (which is clearly Cauchy). Thus, instead of the union of X with the formal limits of limit-less Cauchy sequences, we expect a completion of X to have the underlying set

$$\hat{X} = \{[\{x_m\}] \mid \{x_m\}_{m\geq 1} \text{ is a Cauchy sequence in } X\}$$

where now $[\{x_m\}]$ denotes an equivalence class for a suitable equivalence relation on the set of all Cauchy sequences in X. The following steps realize this idea.

1. Let $\tilde{X}$ be the set of all Cauchy sequences in X and define $\tilde{d} : \tilde{X} \times \tilde{X} \to \mathbb{R}$ by

$$\tilde{d}(\{x_m\}, \{y_m\}) = \lim_{m\to\infty} d(x_m, y_m).$$

2. Prove that the limit defining $\tilde{d}((x_m), (y_m))$ always exists. (Hint: The metric function $d : X \times X \to \mathbb{R}_+$ is uniformly continuous, and $\mathbb{R}$ is complete.)
3. Prove that $(\tilde{X}, \tilde{d})$ is a semimetric space.
4. Recall Exercise 4.4 and let $\hat{X}$ be the metric space thus obtained from $\tilde{X}$.

5. Prove that $(\hat{X}, d)$ is complete. (Hint: Avoid the temptation of taking the diagonal of a Cauchy sequence of Cauchy sequences. Instead, use the Cauchy condition whenever you can.)
6. Note that for all $x \in X$, the constant sequence $\{x_m = x\}_{m \geq 1}$ is Cauchy, thereby defining a function $\iota : X \to \hat{X}$. Prove that it is an isometry.
7. Prove that $\iota(X)$ is dense in $\hat{X}$.

Remark 4.8 Any metric space X has a completion, and thus, if non-empty, X has infinitely many different (yet isomeric) completions. It is important to realize that there is no distinguished completion that deserves to be called *the* completion of X. Any particular construction of a completion for X can thus be seen as a model of the completion of X, whatever it may mean. Since isometric metric spaces are essentially the same it does not matter which model is used for the completion. However, some models may lend themselves to establish certain properties more easily than other models would.

Exercises

Exercise 4.31 For $k \geq 1$ and $I = [a, b]$ a non-degenerate closed interval, show that the metric space $C^k(I, \mathbb{R})$ of functions with a continuous k-th derivative, with the metric induced by the L_∞, norm is not complete.

Exercise 4.32 Is a subspace of a complete metric space necessarily complete?

Exercise 4.33 Prove that a subset of a nowhere dense set is nowhere dense and that a finite union of nowhere dense sets is nowhere dense.

Exercise 4.34 Prove that a subset of a meager set is meager and that a countable union of meager sets is meager. Is an arbitrary union of meager sets necessarily meager?

Exercise 4.35 Is $\mathbb{R} - \mathbb{Q}$, the set of irrational numbers, a meager subset of $\mathbb{R}$ with respect to the Euclidean metric?

Exercise 4.36 Consider the topological space $\mathbb{Q}$ as a subspace of $\mathbb{R}$ endowed with the Euclidean topology. Prove that $\mathbb{Q}$ is metrizable but not by any complete metric on $\mathbb{Q}$.

Exercise 4.37 Let $\{f_m : \mathbb{R} \to \mathbb{R}\}_{m \geq 1}$ be a countable family of continuous functions (with respect to the Euclidean topology on $\mathbb{R}$). Suppose that for every point $x \in \mathbb{R}$ there exist distinct functions f_k, f_m such that $f_k(x) = f_m(x)$. Prove that there exists an interval (a, b), with $a < b$, and distinct functions f_k, f_m such that $f_k(x) = f_m(x)$ for all $x \in (a, b)$.

Exercise 4.38 Let X and Y be metric spaces and consider respective completions $\iota_X : X \to \hat{X}$ and $\iota_Y : Y \to \hat{Y}$. Recall that the product of metric spaces can be metrized by either the Euclidean product metric or the additive product metric. Check, in each case, if the function $\iota : X \times Y \to \hat{X} \times \hat{Y}$, given by $\iota(x, y) = (\iota_X(x), \iota_Y(y))$, is a completion of $X \times Y$.

Exercise 4.39 Let V be an inner product space considered as a metric space and let $\iota : V \to \hat{V}$ be a completion of it. Prove that the inner product on V extends to an inner product on $\hat{V}$.

Exercise 4.40 Let V be a normed space with its induced metric structure, and let $\iota : (V, d) \to (\hat{V}, \hat{d})$ be a completion of it. Prove that the norm $\| - \| : V \to \mathbb{R}$ extends to a norm on $\hat{V}$ which induces the metric $\hat{d}$.

4.5 Compactness and Boundedness

Compactness is a topological property and since, as we saw, any metric space is naturally endowed with a topology, compactness is meaningful in any metric space. The notion of completeness on the other hand is a metric property and does not have a topological counter-part. Completeness and compactness are not independent properties. The space $\mathbb{R}$ (with the Euclidean metric) is complete but not compact, its subset $[0, 1]$ is both complete and compact, and $(0, 1)$ is neither complete nor compact. The final possibility though, as we show below, is implied; if a metric space is compact, then it must be complete. In this section we show the remarkable fact that compactness in a metric space is equivalent to completeness together with a property known as total boundedness. The latter is a purely metric property and thus the result shows that the topological property of compactness is equivalent (for metric spaces, of course) to the conjunction of the purely metric conditions of completeness and total boundedness.

The reader is familiar with the notion of bounded sets in the Euclidean space $\mathbb{R}^n$. In a general metric space though there are two notions of boundedness, given next. These two notions coincide in the Euclidean spaces. Recall that for a subset $S \subseteq X$ of a metric space X, its diameter is given by $\text{diam}(S) = \sup_{x,y \in S}\{d(x, y)\}$, which may be ∞.

Definition 4.17 A subset S in a metric space X is said to be *bounded* if its diameter is finite. A subset S is *totally bounded* if for every $\varepsilon > 0$ one can cover S by finitely many sets of diameter smaller than ε.

The reader is invited to show that every totally bounded set is bounded, but that the converse need not hold. It is also evident that if $S' \subseteq S \subseteq X$ and S is bounded (respectively totally bounded), then S' is bounded (respectively totally bounded). Since X is a subset of itself, we may speak of the metric space X as being bounded or totally bounded.

The following result, whose proof is left for the reader, has an important corollary which is pivotal for the proof of the main theorem below.

Proposition 4.3 *The equality* $\text{diam}(S) = \text{diam}(\overline{S})$ *holds for all subsets* S *of a metric space* X.

Corollary 4.9 *Let X be a totally bounded metric space, $\mathscr{U} = \{U_i\}_{i\in I}$ an open covering of X, and $\varepsilon > 0$. If X can not be covered by finitely many elements from $\mathscr{U}$, then there exists a non-empty closed set $F \subseteq X$ with $\operatorname{diam}(F) < \varepsilon$ that also can not be covered by finitely many elements from $\mathscr{U}$.*

Proof Since X is totally bounded it can be covered by finitely many subsets of diameter smaller than ε. By the preceding proposition, we may take the closures of these sets without altering their diameters and thus we may assume they are all closed. If each of these sets was covered by finitely many elements from $\mathscr{U}$, then (by taking the union) X itself would be covered by finitely many elements from $\mathscr{U}$, contrary to the assumption on X. Thus at least one of these closed sets is the desired one. □

Theorem 4.12 *For a metric space X the following conditions are equivalent.*

1. *X is compact.*
2. *X is sequentially compact, i.e., every sequence $\{x_m\}_{m\geq 1}$ admits a convergent subsequence $\{x_{m_k}\}_{k\geq 1}$.*
3. *X is complete and totally bounded.*

Proof We prove first that if X is compact, then X is sequentially compact, and so let $\{x_m\}_{m\geq 1}$ be a sequence in X. If the set $S = \{x_m \mid m \in \mathbb{N}\}$ is finite, then some of its elements must repeat infinitely often in the given sequence. In other words, the sequence contains a constant subsequence, which certainly converges. Otherwise S is an infinite subset in a compact topological space and thus, by Theorem 3.17, it admits a cluster point $x_0 \in X$ and by Corollary 4.2 this cluster point is the limit of a sequence of elements from S.

Next we show that if X is sequentially compact, then it is complete and totally bounded. Suppose thus that $\{x_m\}_{m\geq 1}$ is a Cauchy sequence in X. By sequential compactness we may extract a subsequence $\{x_{m_k}\}_{k\geq 1}$ converging to a point $x_0 \in X$. We leave it to the reader to show that $x_m \to x_0$ too, and thus every Cauchy sequence converges, so X is complete.

To see that X is also totally bounded, assume for the contrary that it is not. Then there exists an $\varepsilon_0 > 0$ such that

$$X \neq \bigcup_{k=1}^{m} B_{\varepsilon_0}(x_k)$$

for any choice of points $x_1, \ldots, x_m \in X$, $m \geq 1$. We now construct the following sequence. Since $X \neq \emptyset$ (otherwise it is totally bounded), choose an arbitrary element $x_1 \in X$. Suppose that we had chosen points $x_1, \ldots, x_m \in X$ with the property that $d(x_i, x_j) \geq \varepsilon_0$ for all $1 \leq i \neq j \leq m$. We can now choose an arbitrary element

$$x_{m+1} \in X - \bigcup_{k=1}^{m} B_{\varepsilon_0}(k_k)$$

and the condition $d(x_i, x_j) \geq \varepsilon_0$ now holds for all $1 \leq i \neq j \leq m+1$. In this way we obtain the infinite sequence $\{x_m\}_{m\geq 1}$ which, by sequential compactness, has a convergent subsequence, which thus must be Cauchy. But by the construction of $\{x_m\}_{m\geq 1}$, no subsequence of it is Cauchy, and thus we obtain a contradiction.

Finally, we prove that if X is complete and totally bounded, then X is compact. Suppose that $\mathscr{U}$ is an open covering of X that does not admit a finite sub-covering. Our aim will be to construct a non-convergent Cauchy sequence and thus obtain a contradiction.

Let us call a non-empty subset $F \subseteq X$ a *problematic* set if it is closed and can not be covered by finitely many elements from $\mathscr{U}$ (so X itself is problematic). By Corollary 4.9 (applied to X with $\varepsilon = 1$) there exists a problematic set F_1 with $\mathrm{diam}(F_1) < 1$. Applying the same corollary again (to F_1 and with $\varepsilon = 1/2$) there exists a problematic set $F_2 \subseteq F_1$ with $\mathrm{diam}(F_2) < 1/2$. Continuing inductively in this manner we obtain a sequence $F_1 \supseteq F_2 \supseteq F_3 \supseteq \cdots \supseteq F_m \supseteq \cdots$ of non-empty closed sets with $\mathrm{diam}(F_m) < 1/m$, all of which are problematic. Let $x_m \in F_m$ be an arbitrary element in F_m. The sequence $\{x_m\}_{m\geq 1}$ is clearly Cauchy since $x_k \in F_m$ for all $k \geq m$ and thus $d(x_k, x_{k'}) < 1/m$ for all $k, k' > m$. To see that the sequence can not converge assume that $x_k \to x_0$ for some $x_0 \in X$. Recalling that every closed set contains its limit points we see that $x_0 \in F_m$ for all $m \geq 1$ (since $x_k \in F_m$ for all $k \geq m$). Further, since $\mathscr{U}$ covers X there exists an element $U \in \mathscr{U}$ with $x_0 \in U$. Since U is open there exists a $\delta > 0$ with $B_\delta(x) \subseteq U$. Let m now be large enough so that $1/m < \delta$. We then have that $x_0 \in F_m$ and $\mathrm{diam}(F_m) < \delta$ and thus $F_m \subseteq B_\delta(x_0) \subseteq U$. We thus found an element in $\mathscr{U}$ which covers F_m, contrary to F_m being a problematic set. It is thus shown that $\{x_m\}_{m\geq 1}$ is a non-convergent Cauchy sequence, contradicting the completeness of X. The proof is complete. □

Remark 4.9 We remark again that this result expresses the topological property of compactness purely in terms of metric notions.

Exercises

Exercise 4.41 Prove that a totally bounded subset S in a metric space X is bounded.

Exercise 4.42 Let (X, d) be a metric space and let $d_t : X \times X \to \mathbb{R}_+$ be given by $d_t(x, y) = d(x, y)$ if $d(x, y) < 1$ and $d_t(x, y) = 1$ otherwise. Prove that (X, d_t) is a metric space and that both d_t and d induce the same topology on X.

Exercise 4.43 Continuing the preceding exercise, prove that in (X, d_t) every subset is bounded while the totally bounded subsets in (X, d) and (X, d_t) coincide.

Exercise 4.44 Prove that if $\{x_m\}_{m\geq 1}$ is a Cauchy sequence in a metric space X and $\{x_{m_k}\}_{k\geq 1}$ is a convergent subsequence converging to x_0, then $x_m \to x_0$.

Exercise 4.45 Prove that a metric subspace S of $\mathbb{R}^n$ with the Euclidean metric is complete if, and only if, S is closed in $\mathbb{R}^n$.

Exercise 4.46 In $\mathbb{R}^n$ with the Euclidean metric, prove that a subset is bounded if, and only if, it is totally bounded. Conclude that the compact subsets are precisely the closed and bounded ones.

Exercise 4.47 Prove Proposition 4.3.

Exercise 4.48 Prove that any subset of a totally bounded metric space X is totally bounded and that a closed subset of a complete metric space is complete. Deduce that a closed subset of a compact metric space is compact.

Exercise 4.49 Prove that the unit ball $\{x \in \mathbb{R}^n \mid \|x\| \leq 1\}$ is compact in $\mathbb{R}^n$ (here the norm and the topology are the Euclidean ones).

Exercise 4.50 Prove that the unit ball $\{x \in c_0 \mid \|x\|_\infty \leq 1\}$ is not compact in c_0 with the induced topology (c_0 is the space of sequences, say of real numbers, converging to 0 and the norm is the ℓ_∞ norm).

Further Reading

The adventurous reader is urged to delve into Fréchet's original thesis ([3]) for a fascinating glimpse into the mind of the father of the subject matter of this chapter. Of the numerous introductory level textbooks on the subject, the reader may wish to consult [5] for an introduction to metric spaces which is very much in line with the presentation as given in this chapter, or the more advanced [6] for a more comprehensive introduction to analysis in general. The reader intrigued by Baire's Theorem will find a historical account and numerous examples of its uses in [4]

For a glimpse of the overwhelming ubiquity of metric spaces in mathematics see the Encyclopedia of Distances ([1]). The encyclopedia presents an almost uncountable list of examples of metric spaces as well as various generalizations of metric spaces. Some of the generalizations (e.g., [8]) are inspired by problems in physics. Other generalizations (e.g., [2]) give rise to a unifying perspective on the relationship between metric spaces and topology (see [9]). For an investigation in the opposite direction, i.e., metric spaces with special properties, see [7].

References

1. M.M. Deza, *Encyclopedia of Distances*, 2nd edn. (Springer, Heidelberg, 2013)
2. R.C. Flagg, Quantales and continuity spaces. Algebra Universalis **37**(3), 257–276 (1997)
3. M.R. Fréchet, Sur Quelque Points du Calcul Fonctionnel. Rendic. Circ. Mat. **22**, 1–74 (1906)
4. S. Hawtrey, Jones applications of the baire category theorem. real Anal. Exch. **23**(2), 363–394 (1999)
5. M. O'Searcoid, *Elements of Abstract Analysis*, Springer Undergraduate Mathematics Series (Springer, London, 2002)
6. M. O'Searcoid, *Metric Spaces*, Springer Undergraduate Mathematics Series (Springer, London, 2007)
7. R. Rammal, G. Toulouse, M.A. Virasoro, Ultrametricity for physicists. Rev. Mod. Phys. **58**(3), 765–788 (1986)
8. J. Skakala, M. Visser, Bi-metric pseudo-Finslerian spacetimes. J. Geom. Phys. **61**(8), 1396–1400 (2011)
9. I. Weiss, A note on the metrizability of spaces, Algebra Universalis, to appear. http://arxiv.org/abs/1311.4940

Chapter 5
Normed Spaces

Abstract Normed spaces are treated at length as well as techniques of Banach spaces for solving differential and integral equations. Classical results, such as the Open Mapping Theorem, the Closed Graph Theorem, the Hahn-Banach Theorem, the Riesz Representation Theorem, and a few more, are given as well, establishing the core of the theory of bounded operators on Banach spaces. The chapter closes with a view towards generalizations of the theory in two directions, providing a glimpse of the theory of unbounded operators and of locally convex spaces.

Keywords Normed space · Semi-normed space · Banach space · Fixed-point techniques · Dual space · Closed operator · Bounded operator · Linear functional · Fredholm equation · Hahn-Banach Theorem

Normed spaces were introduced in Chap. 2. We revisit the definition here in a broader context and we study the topology and metric structure of normed spaces. We start though with the following discussion.

From elementary geometry of the plane, namely Pythagoras' Theorem, the length, or *norm*, of a vector $x = (x_1, x_2) \in \mathbb{R}^2$ is given by $\sqrt{x_1^2 + x_2^2}$. Consider now two vectors $x, y \in \mathbb{R}^2$. These vectors then determine a triangle with side lengths $\|x\|$, $\|y\|$, and $\|x - y\|$. Let θ be the internal triangle angle formed by x and y. From elementary trigonometry, namely the law of cosines (which is a generalization of Pythagoras' Theorem), the lengths of the sides of the triangle and the angle θ are related by the formula

$$\|x - y\|^2 = \|x\|^2 + \|y\|^2 - 2\|x\|\|y\| \cos \theta.$$

Using the above expression for the length of a vector, this formula becomes

$$(x_1 - y_1)^2 + (x_2 - y_2)^2 = x_1^2 + x_2^2 + y_1^2 + y_2^2 - 2\|x\|\|y\| \cos \theta,$$

which simplifies to

$$x_1 y_1 + x_2 y_2 = \|x\|\|y\| \cos \theta.$$

Recalling the standard inner product in $\mathbb{R}^2$, given by $\langle x, y \rangle = x_1 y_1 + x_2 y_2$, we obtain

© Springer International Publishing Switzerland 2015

C. Alabiso and I. Weiss, *A Primer on Hilbert Space Theory*,
UNITEXT for Physics, DOI 10.1007/978-3-319-03713-4_5

$$\langle x, y\rangle = \|x\|\|y\| \cos\theta$$

and thus we have discovered the standard inner product and the Cauchy-Schwarz Inequality in $\mathbb{R}^2$, by elementary geometry.

We thus see that with a sufficiently strong geometry (i.e., the presence of angles), a notion of a norm gives rise to an inner product. Conversely, an inner product gives rise to a norm by means of $\|x\| = \sqrt{\langle x, x\rangle}$, and, via the Cauchy-Schwarz Inequality, also gives rise to angles. Quite often though, the geometry present in a linear space of interest is not so rich that one can speak of angles, but a notion of a norm is still available. These are the normed spaces, which form the topic of this chapter.

Section 5.1 presents the definition of semi-normed, normed, and Banach spaces, establishes several fundamental continuity results in general normed spaces, and establishes the Open Mapping Theorem, a result of great importance. Section 5.2 then is an investigation of a powerful technique for solving a problem by recasting it in the form of a fixed-point problem, and then resorting to Banach's Fixed-Point Theorem. This technique is exemplified in detail for the solution of certain classes of differential and integral equations, namely to Volterra equations and to Fredholm equations. Section 5.3 starts off with a general study of inverse operators, motivated by an analysis of fixed-point problems, and then proceeds to apply this more powerful technique to strengthen the results obtained in the previous section. Section 5.4 investigates dual spaces, establishes the Riesz Representation Theorem, presents the duals of ℓ_p spaces, and the Hahn-Banach Theorem. The first four sections represent the core material of the theory of bounded linear operators between normed spaces. Section 5.5 is a short introduction to two fruitful directions in which the theory can be generalized, namely by considering unbounded operators between normed spaces and by introducing locally convex spaces.

5.1 Semi-Norms, Norms, and Banach Spaces

Banach spaces are normed spaces in which the algebra and geometry mesh very well together, giving rise to powerful theorems. We introduce Banach spaces in a broad context, starting with general semi-normed and normed spaces, a basic investigation of the topology of normed spaces, followed by a study of bounded linear operators and the Open Mapping Theorem.

5.1.1 Semi-Norms and Norms

The notion of a norm on a linear space is an association of lengths to vectors in such a way that the linear structure is taken into consideration. In practice, for various reasons, it is convenient to introduce a weaker structure, that of a semi-norm on a linear space. Every semi-normed space can canonically be turned into a normed one, and thus one may think of semi-norms as an intermediate step towards obtaining norms.

Definition 5.1 Let V be a linear space over $K = \mathbb{R}$ or over $K = \mathbb{C}$. A *semi-norm* on V is a function associating with every vector $x \in V$ a non-negative real number $p(x)$, the *norm* of x, for which the following conditions hold.

1. $p(\alpha x) = |\alpha| p(x)$, for all $\alpha \in K$ and $x \in V$.
2. The *triangle inequality* holds, i.e., $p(x + y) \leq p(x) + p(y)$, for all $x, y \in V$.

If, moreover, $p(x) = 0$ implies $x = 0$, for all $x \in V$, then the function is called a *norm* on V, in which case we write $\|x\| = p(x)$. A linear space V together with a semi-norm on it is called a *semi-normed space*, and a linear space V together with a norm on it is called a *normed space*. It is common to refer to a (semi-)normed space V, leaving $p(x)$, or $\|x\|$, implicit.

Example 5.1 Numerous examples of normed spaces were already presented in Chap. 2, so we only consider here semi-norms. Firstly, and quite trivially, any linear space V supports the *trivial* semi-norm given by $p(x) = 0$ for all $x \in V$. Less trivially, consider the linear space $C(\mathbb{R}, \mathbb{R})$ of all continuous functions $x : \mathbb{R} \to \mathbb{R}$. The *evaluation* semi-norm is given by $\|x\| = |x(0)|$. Indeed

$$\|\alpha x\| = |\alpha x(0)| = |\alpha||x(0)| = |\alpha|\|x\|$$

and

$$\|x + y\| = |x(0) + y(0)| \leq |x(0)| + |y(0)| = \|x\| + \|y\|.$$

Note that it is obvious that $\|x\| = 0$ need not imply $x = 0$, only that $x(0) = 0$, so this semi-norm is not a norm. This construction is a typical one giving rise to semi-norms by ignoring some of the information embodied in the vectors (in this case, only $x(0)$ is important for determining the semi-norm, all the other values of the function are ignored). Naturally, one may evaluate at any point, not just at $x = 0$, and obtain a family of semi-norms.

We mention here that there is another natural way to obtain semi-norms. The space $C([a, b], \mathbb{R})$ is a space of continuous functions. If one wishes to consider non-continuous functions and to utilize the integral in order to obtain a norm similar in spirit to the L_∞ or L_p norms on $C([a, b], \mathbb{R})$, then one encounters the following difficulty. It is well-known that for a non-continuous function x, it is possible that $\int dt\ |x(t)| = 0$ while $x \neq 0$. For instance, if the function is constantly 0 except at finitely many points. Thus, by allowing non-continuous functions into our linear spaces, the insensitivity of the integral to finite changes in the integrand implies that norms on such spaces are rare. Semi-norms on such spaces though are available.

We will also see below one more instance where semi-norms arise naturally from norms, namely when transferring a norm from a space to a quotient of it.

We now describe the process of turning a semi-normed space into a normed one. This, of course, can not be achieved without paying a price. In this case all non-zero vectors of norm 0 must be quotiented out of existence. This is a small price to pay since vectors of norm 0 are, in a sense, negligible. As stated above, this is an important method for constructing normed spaces, since in practice it is common to have a natural choice of a semi-norm on a given space, and then one passes to the suitable quotient as we now describe.

Theorem 5.1 *Let V be a semi-normed space.*

1. *The set $U = \{x \in V \mid p(x) = 0\}$ is a linear subspace of V.*
2. *The function*

$$\|x + U\| = p(x)$$

is well-defined on the quotient space V/U and with it V/U is a normed space.

Proof

1. In a semi-normed space we have $p(0) = p(0 \cdot 0) = |0| \cdot p(0) = 0$, and thus $0 \in U$, which is thus in particular non-empty. It remains to show that U is closed under addition and scalar products. Indeed, if $x, y \in U$, then

$$0 \leq p(x + y) \leq p(x) + p(y) = 0 + 0 = 0,$$

and thus $p(x + y) = 0$, so that $x + y \in U$. Finally, if $x \in U$ and $\alpha \in K$ is an arbitrary scalar, then $p(\alpha x) = |\alpha| p(x) = |\alpha| \cdot 0 = 0$, so that $\alpha x \in U$.

2. To show that $x + U \mapsto p(x)$ is well-defined, suppose that $x' + U = x + U$. Then $x - x' \in U$, namely $p(x - x') = 0$, and thus

$$\|x + U\| = p(x) = p((x - x') + x') \leq p(x - x') + p(x') = p(x') = \|x' + U\|.$$

A similar argument shows that $\|x' + U\| \leq \|x + U\|$, and it thus follows that $\|x + U\| = \|x' + U\|$ and thus that $x + U \mapsto p(x)$ is independent of the choice of representative. Next, clearly, $\|x + U\| \geq 0$ for all $x + U \in V/U$ and

$$\|\alpha(x + U)\| = \|\alpha x + U\| = p(\alpha x) = |\alpha| p(x) = |\alpha| \|x + U\|$$

for all $\alpha \in K$ and $x+U \in V/U$. A similar argument proves the triangle inequality. It remains to show that if $\|x + U\| = 0$, then $x + U = 0$ in the quotient space, namely that $x \in U$. Indeed, since $\|x + U\| = p(x)$, if $\|x + U\| = 0$, then $x \in U$ by definition of U. To conclude, $p(x)$ induces a norm on the quotient space V/U, as claimed. □

We already proved that a normed space is automatically endowed with a distance function. We now extend this simple, yet pivotal, observation to make a connection between semi-normed spaces and semimetric spaces as well. Recall that a semimetric space is a set together with a distance function $d : X \times X \to \mathbb{R}_+$ satisfying the axioms of a metric space, except that it is possible that $d(x, y) = 0$ for distinct points x and y. For the reader's convenience we repeat the statement and the proof for normed spaces as well.

Theorem 5.2 *Let V be a semi-normed space. Then defining $d(x, y) = p(x - y)$, for all vectors $x, y \in V$, endows V with the structure of a semimetric space. If V is a normed space, then $d(x, y) = p(x - y) = \|x - y\|$ endows V with the structure of a metric space. These are, respectively, the* induced *semimetric and metric structures.*

Proof Clearly $d(x, y) \geq 0$. Symmetry follows by

$$d(x, y) = p(x - y) = p((-1)(y - x)) = |-1|p(y - x) = d(y, x)$$

while the triangle inequality follows by

$$d(x, z) = p(x - z) = p(x - y + y - z) \leq p(x - y) + p(y - z) = d(x, y) + d(y, z).$$

Finally, if the semi-norm is in fact a norm, then

$$d(x, y) = 0 \implies \|x - y\| = 0 \implies x - y = 0 \implies x = y,$$

completing the proof. □

Recall that there is a canonical construction for turning a semimetric space into a metric space. In more detail, if (X, d) is a semimetric space, then defining $x \sim y$ precisely when $d(x, y) = 0$, is an equivalence relation on X and the function $d([x], [y]) = d(x, y)$ is well-defined and turns the quotient set $X/\sim$ into a metric space. The next result shows that starting with a semi-normed space, turning it into a metric space by first constructing the associated semimetric and then turning it into a metric space, or first turning it into a normed space and then associating a metric space with it, yields the same end-result.

Theorem 5.3 *Let V be a semi-normed space. Consider the induced normed space $V/U = \{x + U \mid x \in V\}$, where $U = \{x \in V \mid p(x) = 0\}$, and the induced metric space $(V/U, d)$. Consider next the induced semimetric space (V, d) and the induced metric space $(V/\sim, d)$ where $\sim$ is the equivalence relation on V given by $x \sim y$ precisely when $d(x, y) = 0$. Then the two resulting metric spaces are identical.*

Proof First we show that the two resulting metric spaces have identical underlying sets. Indeed, for the second described metric space we consider the quotient set $V/\sim$ where $x \sim y$ if, and only if, $d(x, y) = 0$. But $d(x, y) = p(x - y)$ and thus $x \sim y$ if, and only if, $x - y \in U$. It follows that $[x] = x + U$, and thus the quotient set $V/\sim$ coincides with the quotient set construction for V/U. Next we show that the metrics agree as well. The induced metric on V/U is given by

$$d(x + U, y + U) = \|x - y + U\| = p(x - y)$$

while the metric d on $V/\sim$ is given in terms of the semimetric d on V as follows:

$$d([x], [y]) = d(x, y) = p(x - y).$$

The two metrics are thus identical, as claimed. □

A subspace U of a semi-normed space (V, p) is automatically a semi-normed space by defining $y \mapsto p(y)$ for every $y \in U$, using the semi-norm in the ambient space. Similarly, any subspace U of a normed space $(V, \|\cdot\|)$ is a normed space with norm given by $y \mapsto \|y\|$, for all $y \in U$. The proofs are immediate. The situation for quotient spaces is slightly more complicated and we now address it. At this point topological considerations enter the discussion, and thus we recall that any metric induces a topology and thus the concepts of topology are available in any normed space. In particular, we may speak of closed subsets of a normed space V.

Theorem 5.4 *Let V be a normed space and U a linear subspace of V. Then the quotient space V/U, when endowed with the function*

$$p(x+U) = \inf_{u\in U} \|x+u\|,$$

is a semi-normed space. If U is a closed subspace of V, then V/U is in fact a normed space.

Proof The quotient vector space V/U is the set $\{x+U \mid x \in V\}$ of translations of U, with the linear operations determined by the representatives. We need to verify that the proposed formula

$$p(x+U) = \inf_{u\in U} \|x+u\|$$

satisfies the conditions for being a semi-norm. Clearly, $p(x+U) \geq 0$ since it is an infimum of non-negative real numbers. Given $\alpha \in K$ and $x+U \in V/U$ we need to show that

$$p(\alpha x+U) = |\alpha| p(x+U).$$

If $\alpha = 0$, then

$$p(\alpha x+U) = p(0+U) = \inf_{u\in U} \|u\| = 0$$

(since $0 \in U$ and $\|0\| = 0$), and the equality holds. If $\alpha \neq 0$, then

$$p(\alpha x+U) = \inf_{u\in U} \|\alpha x+u\| = \inf_{u\in U} \|\alpha x+\alpha\frac{1}{\alpha}u\| = |\alpha| \inf_{u\in U} \|x+\frac{1}{\alpha}u\|$$

and noting that $\{(1/\alpha)u\}_{u\in U}$ and $\{u\}_{u\in U}$ are the same set of vectors, the desired equality follows. To establish the triangle inequality we need to show that

$$\inf_{u\in U} \|x+y+u\| \leq \inf_{u'\in U} \|x+u'\| + \inf_{u''\in U} \|y+u''\| = \inf_{u',u''\in U} \|x+u'\| + \|y+u''\|.$$

It suffices to establish that, given $u', u'' \in U$,

$$\inf_{u \in U} \|x + y + u\| \leq \|x + u'\| + \|y + u''\|,$$

and indeed,

$$\inf_{u \in U} \|x+y+u\| \leq \|x+y+(u'+u'')\| = \|(x+u')+(y+u'')\| \leq \|x+u'\|+\|y+u''\|.$$

We thus far established that V/U, with the proposed function, is a semi-normed space. To complete the proof we now show that if U is closed in V, then V/U is in fact a normed space, namely that for $\|x + U\| = p(x + U)$, if $\|x + U\| = 0$, then $x + U = 0$ in the quotient space, that is that $x \in U$. So, noting another equivalent form for the proposed norm, suppose that

$$\|x + U\| = \inf_{u \in U} \|u - x\| = 0.$$

Now, by definition of infimum, there exists a sequence $u_n \in U$ with $\|u_n - x\| \to 0$. In other words, $d(u_n, x) \to 0$ and thus $u_n \to x$ in the induced topology. We thus exhibited x as a limit of a sequence in the closed set U, and thus x itself belongs to U, as required. □

To see that if U is not closed then the quotient space may indeed only be a semi-normed space, consider any normed space V and a dense linear subspace U of it. The quotient space V/U with the induced semi-norm

$$p(x + U) = \inf_{u \in U} \|x + u\| = \inf_{u \in U} \|u - x\|$$

allows for $p(x+U) = 0$ even if $x \notin U$. In fact, $p(x+U) = 0$ for all $x \in V$, in other words the induced semi-norm on V/U degenerates to the trivial semi-norm. Indeed, given $x \in V$, choose a sequence in U that converges to x, namely $\|u_n - x\| \to 0$. In particular then,

$$p(x + U) \leq \inf_{n} \|u_n - x\| = 0.$$

5.1.2 Banach Spaces

It can be shown that any finite-dimensional normed space is complete. However, as we have seen, an infinite-dimensional normed linear space need not be complete. Non-complete normed spaces exhibit very pathological behaviour. In some sense, such spaces are full of holes. In the rest of this chapter we will explore some of the consequence of the powerful interaction between algebra and topology when a normed space is also a Banach space, namely a complete metric space.

Definition 5.2 A normed space which, with the metric induced by the norm, is complete is called a *Banach space*. Banach spaces are usually denoted by $\mathscr{B}$. An inner product space which, with the metric induced by the inner product, is complete is called a *Hilbert space*. Hilbert spaces are typically denoted by $\mathscr{H}$.

Example 5.2 We have shown that $\mathbb{R}^n$ with the Euclidean metric is complete. Since the Euclidean metric is induced by the standard inner product on $\mathbb{R}^n$, it follows that $\mathbb{R}^n$ is a Hilbert space. For all $1 \leq p \leq \infty$ the normed space ℓ_p was shown to be complete with respect to the induced metric, and thus ℓ_p is a Banach space. The space ℓ_2 is a Hilbert space since the ℓ_2 norm is induced by an inner product. The space $C([a, b], \mathbb{R})$ with the L_∞ norm is complete and is thus a Banach space. However, the pre-L_p spaces $C([a, b], \mathbb{R})$ with the L_p norm for $1 \leq p < \infty$ are not Banach spaces since they are not complete.

Since any metric space has a completion, it is natural to consider the completion of the non-complete spaces $C([a, b], \mathbb{R})$ (with various norms) in the hope that the algebraic structure of the space can be extended to the completion so as to obtain a Banach space. In fact, Exercises 4.39 and 4.40 already accomplish just that, a fact we record as the following result.

Theorem 5.5 *Let V be a linear space.*

1. *Given an inner product on V, let $\hat{V}$ be a completion of V with respect to the metric induced by the norm (induced by the inner product). Then the inner product on V extends to an inner product on $\hat{V}$, making $\hat{V}$ a Hilbert space.*
2. *Given a norm on V, let $\hat{V}$ be a completion of V with respect to the metric induced by the norm. Then the norm on V extends to a norm on $\hat{V}$, making $\hat{V}$ a Banach space.*

Definition 5.3 Let $1 \leq p < \infty$. A completion of $C([a, b], \mathbb{R})$ with respect to the L_p norm is denoted by $L_p([a, b], \mathbb{R})$.

Corollary 5.1 *For all $1 \leq p < \infty$ the space $L_p([a, b], \mathbb{R})$ is a Banach space, and $L_2([a, b], \mathbb{R})$ is a Hilbert space.*

Remark 5.1 This family of Banach spaces is of great importance in applications of the general theory to problems in Quantum Mechanics. It should be noted that a more common route to the definition of the L_p spaces is via measure theory and Lebesgue integrable functions. We took here a short-cut (of sorts) to the definition, utilizing the results on metric spaces established earlier.

Recall that a series $\sum_{k=1}^{\infty} a_k$ of real or complex numbers is said to converge to s if the sequence $\{s_m\}_{m \geq 1}$ of partial sums, i.e., $s_m = \sum_{k=1}^{m} a_k$, converges to s. We thus see that convergence for series is defined in terms of sequences. Moreover, one can recover the convergence of sequences in terms of that of series, as follows. Given a sequence $\{x_m\}_{m \geq 1}$ consider the series $\sum_{k=1}^{\infty} a_k$ where $a_k = x_k - x_{k-1}$ (and $a_1 = x_1$). The partial sum s_m is easily seen to be nothing but x_m, and thus if the series converges, then so does the sequence. Some aspects of the interplay between sequences and series in $\mathbb{R}$ and in $\mathbb{C}$ extend to normed spaces, as we now briefly discuss.

Definition 5.4 For elements $\{a_m\}_{m\geq 1}$ in a normed space V, we say that the *series* $\sum_{k=1}^{\infty} a_k$ *converges* to $s \in V$ if the sequence of partial sums converges to s with respect to the norm in V, i.e., when

$$\lim_{m\to\infty} \|(\sum_{k=1}^{m} a_k) - s\| = 0.$$

We say that a series

$$\sum_{k=1}^{\infty} a_k$$

converges absolutely if the series of real numbers

$$\sum_{k=1}^{\infty} \|a_k\|$$

converges.

Recall that if a series $\sum_{k=1}^{\infty} a_k$ of real (or complex) numbers converges absolutely, that is if $\sum_{k=1}^{\infty} |a_k|$ converges, then the original series converges as well. Viewing $\mathbb{R}$ (or $\mathbb{C}$) as a normed space, this result is a special case of the following useful criterion for a normed space to be a Banach space.

Theorem 5.6 *A normed space V is a Banach space if, and only if, the absolute convergence of a series in V implies its convergence.*

The proof is left for the reader as we now turn to show that the algebraic structure in a normed space interacts well with the metric and topological structures induced by the norm.

Proposition 5.1 *The following assertions hold for any normed space V.*

1. *The mapping $x \mapsto \|x\|$, as a function $V \to \mathbb{R}$, is non-expanding and thus the norm function is (uniformly) continuous.*
2. *For any fixed $x_0 \in V$, the translation mapping $x \mapsto x_0 + x$ is a global isometry and a homeomorphism.*
3. *For any $\alpha \in K$ with $\alpha \neq 0$, the scaling mapping $x \mapsto \alpha x$, as a function $V \to V$, is uniformly continuous and a homeomorphism.*
4. *The mapping $(x, y) \mapsto x + y$, as a function $V \times V \to V$, is continuous (and in fact uniformly continuous when $V \times V$ is given the additive product metric).*

Proof

1. Referring to Proposition 2.12 we have

$$d_{\mathbb{R}}(\|x\|, \|y\|) = |\|x\| - \|y\|| \leq \|x - y\| = d_V(x, y).$$

2. Since $\|(x_0 + x) - (x_0 + x')\| = \|x - x'\|$ any translation is an isometry. Further, the translation $x \mapsto x_0 + x$ is clearly invertible, its inverse being the translation $x \mapsto -x_0 + x$. An isometry is certainly continuous, and thus any translation is a continuous function with continuous inverse, and thus a homeomorphism.
3. Since $\|\alpha x - \alpha x'\| = |\alpha| \|x - x'\|$, it follows that any scaling mapping is uniformly continuous. The scaling mapping $x \mapsto \alpha x$ is clearly invertible, with inverse the scaling mapping $x \mapsto (1/\alpha)x$. Any non-zero scaling is thus a continuous function with a continuous inverse, and thus a homeomorphism.
4. Left for the reader. □

5.1.3 Bounded Operators

In the text below, linear operators will typically be denoted by A or B, and their values on vectors x will be given by, e.g., Ax rather than the more cumbersome $A(x)$.

The norm $\|x\|$ in a normed space represents the length of the vector x. Given an operator A between normed spaces it is thus natural to study the way the operator affects the norm, that is, how $\|x\|$ and $\|Ax\|$ are related. If the operator does not alter the norm of any vector by more than some constant multiple, then it is said to be bounded. This class of operators is studied here, showing that for linear operators the concept of boundedness and of continuity coincide. The main result we establish is that the collection of bounded linear operators between normed spaces is itself a normed space, which in fact is a Banach space if the codomain is a Banach space.

Definition 5.5 An operator $A : V \to W$ between normed spaces is called a *bounded operator* if there exists a positive real number M such that

$$\|Ax\| \leq M\|x\|$$

for all $x \in V$.

Equivalently, an operator A is bounded if it maps bounded sets in the domain to bounded sets in the codomain (where a set S of vectors in a normed space is bounded if there exists a positive M such that $\|s\| \leq M$ for all $s \in S$).

Remark 5.2 It is not hard to show that any linear operator between finite-dimensional normed spaces is bounded.

Example 5.3 In $C([a, b], \mathbb{R})$, the linear space of continuous functions $x : [a, b] \to \mathbb{R}$ with the L_∞ norm, consider the integral operator:

$$(Ax)(s) = \int_a^b dt\ K(t, s)\ x(t),$$

with $K : [a, b] \times [a, b] \to \mathbb{R}$ a given continuous function. To see that the operator A is bounded note that if $M = \max_{t,s\in[a,b]} |K(t, s)|$, then

$$\|Ax\| = \max_{s\in[a,b]} \left| \int_a^b dt\, K(t,s)\, x(t) \right| \le \max_{t,s\in[a,b]} \left|K(t,s)\right| \left| \int_a^b dt\, x(t) \right| \le$$
$$\le M(b-a) \max_{t\in[a,b]} \left|x(t)\right| = M(b-a)\|x\|.$$

The following example shows that a very familiar linear operator is not bounded.

Example 5.4 Consider the operator $A : C^1([0, 1], \mathbb{R}) \to C([0, 1], \mathbb{R})$, where both the domain and the codomain are given the L_∞ norm, which maps any continuously differentiable function $x : [0, 1] \to \mathbb{R}$ to its derivative, i.e., $Ax = dx/dt$. The fact that A is not bounded is seen for example by considering $x_n(t) = t^n$, and noting that $\|x_n\| = 1$ but $Ax_n(t) = nt^{n-1}$, and thus $\|Ax_n\| = n$.

Remark 5.3 The failure of the differentiation operator to be bounded is, in a sense, a reflection (or the cause, depending on one's point-of-view) of the complicated nature of its inverse operator, namely the non-trivial nature of integration.

As mentioned above, it is the case for linear operators that being continuous and being bounded are equivalent conditions. In fact, we now show that the interplay between the linear structure and the topology in a normed space implies a stronger unification of concepts.

Theorem 5.7 *The following conditions for a linear operator $A : V \to W$ between normed spaces are equivalent.*

1. *A is uniformly continuous.*
2. *A is continuous.*
3. *A is continuous at $x = 0$.*
4. *A is bounded.*

Proof Obviously, uniform continuity implies continuity, and continuity implies continuity at $x = 0$. Suppose now that A is continuous at $x = 0$. Then there exists a $\delta > 0$ such that $\|Ax\| \le 1$ for all $x \in V$ with $\|x\| < 2\delta$. Now, given an arbitrary $y \in V$, if $y = 0$, then $Ay = 0$ while if $y \neq 0$, then, noting that $\|\delta y/\|y\|\| = \delta < 2\delta$, one obtains that

$$\|Ay\| = \|A\frac{\delta\|y\|y}{\delta\|y\|}\| = \frac{\|y\|}{\delta}\|A\frac{\delta y}{\|y\|}\| \le \frac{1}{\delta}\|y\|,$$

showing that A is bounded by $1/\delta$. Finally, if A is bounded by some $M > 0$, then $\|Ax - Ax'\| = \|A(x - x')\| \le M\|x - x'\|$, for all $x, x' \in V$, and thus uniform continuity follows easily. □

For a bounded operator A, the inequality $\|Ax\| \leq M\|x\|$ sets an upper bound on how much A can lengthen vectors in the domain. It is natural to consider the infimum over all such upper bounds M and define that to be the norm of the operator A.

Definition 5.6 Let $A : V \to W$ be a linear operator. The *operator norm* or simply the *norm* of A is defined to be

$$\|A\| = \inf\{M \geq 0 \mid \forall x \in V \quad \|Ax\| \leq M\|x\|\},$$

adopting the convention that $\inf \emptyset = \infty$.

Thus, a linear operator is bounded if, and only if, its norm is finite.

The homogeneity of the norm with respect to the scalar product, and continuity considerations, give rise to the following ways of computing the operator norm.

Proposition 5.2 *For a bounded linear operator $A : V \to W$ between normed spaces, the following expressions are all equal to $\|A\|$.*

1. $\sup\{\|Ax\| \mid x \in V, \quad \|x\| \leq 1\}$.
2. $\sup\{\|Ax\| \mid x \in V, \quad \|x\| < 1\}$.
3. $\sup\{\|Ax\| \mid x \in V, \quad \|x\| = 1\}$.
4. $\sup\{\|Ax\|/\|x\| \mid x \in V, \quad x \neq 0\}$.

The proof is left for the reader.

In general, any bounded operator, simply since it is continuous, is determined by its values on a dense subset. However, not any bounded operator on a dense subset can be extended to the full domain. The next result shows that when the codomain is a Banach space this difficulty disappears.

Theorem 5.8 *Let V be a normed space, $\mathscr{D} \subseteq V$ a dense linear subspace of V, and $\mathscr{B}$ a Banach space. If $A : \mathscr{D} \to \mathscr{B}$ is a bounded linear operator, then there exists a unique linear operator $B : V \to \mathscr{B}$ such that $Bx = Ax$ for all $x \in \mathscr{D}$ and $\|B\| = \|A\|$. Such a B is called an extension of A.*

Proof Given $x_0 \in V$ let $\{x_m\}_{m\geq 1}$ be a sequence in $\mathscr{D}$ with $x_m \to x_0$, and thus in particular, $\{x_m\}_{m\geq 1}$ is Cauchy. Recall that a bounded linear operator is in particular a uniformly continuous function, and that uniformly continuous functions preserve Cauchy sequences. Hence A preserves Cauchy sequences, and thus the sequence $\{Ax_m\}_{m\geq 1}$ in $\mathscr{B}$ is Cauchy. Since $\mathscr{B}$ is complete there exists an element $y_0 \in \mathscr{B}$ with $Ax_m \to y_0$. Suppose now that $\{x'_m\}_{m\geq 1}$ is another sequence in $\mathscr{D}$ with $x'_m \to x_0$, and suppose $Ax'_m \to y'_0$. Utilizing the continuity of the norm, it follows that

$$\begin{aligned} \|y_0 - y'_0\| &= \|\lim_{m\to\infty}(Ax_m - Ax'_m)\| = \lim_{m\to\infty}\|Ax_m - Ax'_m\| \leq \\ &\leq \|A\|\|\lim_{m\to\infty}(x_m - x'_m)\| = \|A\|\|x - x\| = 0 \end{aligned}$$

and thus $y_0 = y'_0$. Setting $Bx_0 = y_0$ is thus well-defined and we obtain a function $B : V \to \mathscr{B}$. Obviously, for $x \in \mathscr{D}$ we may choose $x_m = x$ and then

$$Bx = \lim_{m\to\infty} Ax_m = Ax$$

showing that B extends A.

That B is linear is seen as follows. If $x, y \in V$ and $x_m \to x$ and $y_m \to y$ as above, then it follows that $x_m + y_m \to x + y$ and thus

$$B(x+y) = \lim_{m\to\infty} A(x_m + y_m) = \lim_{m\to\infty} Ax_m + \lim_{m\to\infty} Ay_m = Bx + By.$$

A similar argument shows that $B(\alpha x) = \alpha(Bx)$. As for showing that $\|A\| = \|B\|$, note first that $\|A\| \leq \|B\|$ just by the fact that B is an extension of A. Next, for $x \in V$ and $x_m \to x_0$ as above, one has

$$\|Ax_m\| \leq \|A\| \|x_m\|$$

and passing to the limit, and remembering that the norm is continuous, we obtain

$$\|Bx\| \leq \|A\| \|x\|$$

showing that $\|B\| \leq \|A\|$, and thus the desired equality holds. Finally, if C is any bounded linear extension of A, then it is continuous and agrees with A on a dense subset. But any two continuous mappings that agree on a dense subset are equal, hence $B = C$, showing that B is the unique linear bounded extension of A to all of V. □

5.1.4 The Open Mapping Theorem

A continuous function between topological spaces need not be open (i.e., it need not send open subsets to open subsets). We now show that surjectivity is a sufficient condition for a bounded linear operator between Banach spaces to be open. The result is due to Stefan Banach and is our first example of a deep result in the theory of Banach spaces, one with profound consequences.

The technical heart of the argument is stated as a separate lemma. Its proof is omitted in favour of a more conceptual presentation of the the main result. We refer the reader to any text dedicated to Banach spaces in order to fill-in the details and obtain a fully rigorous proof.

It will be convenient to introduce the notation $0_V(s) = \{x \in V \mid \|x\| < s\}$, the open ball in a normed space V with centre the zero vector 0 and radius $s > 0$. We further write $n \cdot 0_V(s)$ for the point-wise scaling of $0_V(s)$ by n, namely

$$n \cdot 0_V(s) = \{nx \mid x \in 0_V(s)\} = 0_V(ns).$$

Lemma 5.1 *Let $A : \mathscr{B} \to V$ be a bounded linear operator from a Banach space $\mathscr{B}$ to a normed space V. It then follows that if $r, s > 0$ are such that*

$$0_V(s) \subseteq \overline{A(0_{\mathscr{B}}(r))},$$

then

$$0_V(s) \subseteq A(0_{\mathscr{B}}(r)).$$

Theorem 5.9 (Open Mapping Theorem) *A bounded surjective linear operator $A : \mathscr{B}_1 \to \mathscr{B}_2$ between Banach spaces is an open mapping.*

Proof Let $U \subseteq \mathscr{B}_1$ be an open set, which we may assume is non-empty. For some $x_0 \in U$, let us consider Ax_0, an arbitrary element in $A(U)$. As U is open, let $\varepsilon > 0$ be such that $B_\varepsilon(x_0) \subseteq U$, and notice that $B_\varepsilon(x_0) = 0_{\mathscr{B}_1}(\varepsilon) + x_0$, i.e., a ball with centre x_0 is the x_0 translation of a ball centered at the origin. By linearity of A we then have

$$A(U) \supseteq A(x_0 + 0_{\mathscr{B}_1}(\varepsilon)) = Ax_0 + A(0_{\mathscr{B}_1}(\varepsilon)).$$

Thus the proof is reduced to showing that for every $\varepsilon > 0$ there exists a $\delta > 0$ such that $0_{\mathscr{B}_2}(\delta) \subseteq A(0_{\mathscr{B}_1}(\varepsilon))$, since then

$$A(U) \supseteq Ax_0 + 0_{\mathscr{B}_2}(\delta) = B_\delta(Ax_0)$$

and thus Ax_0 is an interior point of $A(U)$. As Ax_0 was arbitrary, every point of $A(U)$ is interior, and thus $A(U)$ is open.

We thus fix an open ball $0_{\mathscr{B}_1}(\varepsilon)$, $\varepsilon > 0$, denote $X = A(0_{\mathscr{B}_1}(\varepsilon))$, and we seek out a $\delta > 0$ for which $0_{\mathscr{B}_2}(\delta) \subseteq X$. Noting that

$$\mathscr{B}_1 = \bigcup_{n \geq 1} n \cdot 0_{\mathscr{B}_1}(\varepsilon)$$

the surjectivity and homogeneity of A implies that

$$\mathscr{B}_2 = A(\mathscr{B}_1) = \bigcup_{n \geq 1} n \cdot X.$$

Taking closures and applying Baire's Theorem, there exists an $n \geq 1$ such that $\overline{n \cdot X}$ contains an open ball. In other words, there exists a $\delta > 0$ such that $B_\delta(y) \subseteq \overline{n \cdot X}$. We omit some details on translations and scaling which are used to conclude that in fact from $B_\delta(y) \subseteq \overline{n \cdot X}$ it follows that $0_{\mathscr{B}_2}(\delta/n) \subseteq \overline{X}$. Lemma 5.1 then concludes the proof. □

The next result is an immediate corollary of the Open Mapping Theorem, illustrating some of the taming effect of the completeness requirement on Banach spaces.

Corollary 5.2 *If a bounded linear operator* $A : \mathscr{B}_1 \to \mathscr{B}_2$ *between banach spaces is invertible, then its inverse* A^{-1} *is also a bounded linear operator.*

Proof By Proposition 2.8, the inverse of a linear operator is itself linear. It remains to show that A^{-1} is continuous, namely that $(A^{-1})^{-1}(U)$ is an open set whenever U is open. Since $(A^{-1})^{-1} = A$, we need to show that $A(U)$ is open whenever U is open, namely that A is an open mapping. Indeed, since A is invertible it is necessarily surjective, and thus A is an open mapping by the Open Mapping Theorem. □

5.1.5 Banach Spaces of Linear and Bounded Operators

We end this section by studying the set $\mathbf{B}(V, W)$ of all bounded linear operators $A : V \to W$ between normed spaces. The main result is that when endowed with the operator norm one obtains a normed space, which, furthermore, is a Banach space provided the codomain is a Banach space.

Theorem 5.10 *Let* $A, B : V \to W$ *be bounded linear operators between normed spaces. Then*

1. $\|A\| = 0$ *if, and only if,* $A = 0$.
2. $\|\alpha A\| = |\alpha|\,\|A\|$ *for all scalars* α.
3. $\|A + B\| \leq \|A\| + \|B\|$.

Proof

1. If $A = 0$ then $\|Ax\| \leq 0 \cdot \|x\|$ and thus $\|A\| = 0$. Conversely, if $\|A\| = 0$, then $\|Ax\| \leq \|A\|\|x\| = 0$ and thus $Ax = 0$, for all $x \in V$.
2. $\|\alpha A\| = \sup_{\|x\|=1} \|\alpha Ax\| = |\alpha| \sup_{\|x\|=1} \|Ax\| = |\alpha|\|A\|$.
3. $\|A + B\| = \sup_{\|x\|=1} \|(A + B)x\| \leq \sup_{\|x\|=1}(\|Ax\| + \|Bx\|) = \|A\| + \|B\|$.

□

Recall Theorem 2.3, stating that for linear spaces V and W the set $\mathrm{Hom}(V, W)$ of all linear operators $A : V \to W$ is a linear space.

Theorem 5.11 *For normed spaces* V *and* W*, the set* $\mathbf{B}(V, W)$ *is a linear subspace of* $\mathrm{Hom}(V, W)$ *which, when endowed with the operator norm, is a normed space.*

Proof Theorem 5.10 states that $\mathbf{B}(V, W)$ contains 0 and is closed under addition and scalar multiplication, and is thus a linear subspace of $\mathrm{Hom}(V, W)$. Further, the same theorem establishes the axioms of a normed space for the operator norm on $\mathbf{B}(V, W)$, and the claim follows. □

Since $\mathbf{B}(V, W)$ is a normed space, it is also a metric space and thus a corresponding notion of convergence for operators is automatic.

Definition 5.7 (*Operator Norm Convergence or Strong Convergence*) Let V and W be normed spaces. Given a sequence $\{A_m\}_{m\geq 1}$ of operators in $\mathbf{B}(V, W)$, we say that $\{A_m\}_{m\geq 1}$ *converges strongly* or *converges in the operator norm* to an operator $A_0 \in \mathbf{B}(V, W)$ if $\|A_m - A_0\| \to 0$ (equivalently, when $A_m \to A_0$ with respect to the metric induced by the operator norm).

A weaker notion of convergence is also available in the space $\mathbf{B}(V, W)$.

Definition 5.8 Let V and W be normed spaces. We say that a sequence $\{A_m\}_{m\geq 1}$ of operators in the space $\mathbf{B}(V, W)$ *converges point-wise* to an operator $A_0 \in \mathbf{B}(V, W)$ if $A_m x \to A_0 x$ for all $x \in V$.

Since $\|A_m x - A_0 x\| \leq \|A_m - A_0\| \|x\|$ it follows that strong convergence implies point-wise convergence. That the converse may fail is illustrated next.

Example 5.5 Consider the sequence of linear operators $F_m : \ell_2 \to \mathbb{C}$ (linear operators whose codomain is the ground field are called linear functionals and are studied in detail below), given by $F_m x = x_m$, that is F_m is the projection of the m-th coordinate, and thus is clearly a linear operator. Since

$$\|F_m x\| = |x_m| \leq \|x\|$$

we see that $\|F_m\| \leq 1$ and thus F_m is bounded (and in fact $\|F_m\| = 1$, since there clearly exist $x \in \ell_2$ with $\|F_m x\| = \|x\|$).

We claim that $F_m \to 0$ point-wise, i.e., that $F_m x \to 0$ for all $x \in \ell_2$. Indeed, for $x \in \ell_2$ one has

$$\sum_{m=1}^{\infty} |x_m|^2 < \infty$$

and consequently, $x_m \to 0$. But $x_m = F_m x$, and the claim follows. However, as noted, $\|F_m\| = 1$ and thus $\{F_m\}_{m\geq 1}$ does not converge strongly to 0, or to any other vector.

Finally, we prove the completeness of $\mathbf{B}(V, \mathscr{B})$ when the codomain is complete.

Theorem 5.12 *Let $\mathscr{B}$ be a Banach space and V an arbitrary normed space. Then the linear space $\mathbf{B}(V, \mathscr{B})$, endowed with the operator norm, is a Banach space.*

Proof We already know that with the operator norm $\mathbf{B}(V, \mathscr{B})$ is a normed space, so it remains to show that it is complete. Consider a Cauchy sequence $\{A_m\}_{m\geq 1}$. Since $\|A_k x - A_m x\| \leq \|A_k - A_m\| \|x\| \longrightarrow 0$ for all $x \in V$, it follows at once that $\{A_m x\}_{m\geq 1}$ is also a Cauchy sequence, and, as $\mathscr{B}$ is complete, $A_m x \to y$. We thus define $A_0(x) = y$ and obtain the function $A_0 : V \to \mathscr{B}$ which we proceed to show is a bounded linear operator.

The linearity of A_0 follows by standard limit arguments. For instance, continuity of vector addition implies that

$$A_0(x + x') = \lim_{m\to\infty} A_m(x + x') = \lim_{m\to\infty} A_m x + \lim_{m\to\infty} A_m x' = Ax + Ax',$$

and thus A_0 is additive. A similar argument shows A_0 is homogenous. To show that A_0 is bounded, we use the continuity of the norm to obtain that

$$\|Ax\| = \| \lim_{m\to\infty} A_m x\| = \lim_{m\to\infty} \|A_m x\| \leq \lim_{m\to\infty} \|A_m\| \|x\| \leq M\|x\|$$

where M is any upper bound of the sequence $\|A_m\|$ (which exists since any Cauchy sequence is bounded).

Lastly, we verify that $A_m \to A_0$ in the operator norm. Given $\varepsilon > 0$, using the Cauchy condition, there exists an $N \in \mathbb{N}$ such that $\|A_m - A_k\| < \varepsilon$ for all $m, k > N$. In particular, for all $x \in V$,

$$\|(A_m - A_k)x\| \leq \|A_m - A_k\| \|x\| < \varepsilon\|x\|,$$

for all $m, k \geq N$. When k tends to ∞ we thus obtain

$$\|(A_m - A_0)x\| \leq \varepsilon\|x\|$$

for all $x \in V$ and thus

$$\|A_m - A_0\| \leq \varepsilon$$

for all $m > N$. In other words, $\|A_m - A_0\| \to 0$, as required. □

Exercises

Exercise 5.1 Consider the Euclidean spaces $\mathbb{R}^n$ with the standard inner product, and its associated norm. Prove that any linear operator $A : \mathbb{R}^n \to \mathbb{R}^m$ is bounded.

Exercise 5.2 Prove Proposition 5.2.

Exercise 5.3 Prove Theorem 5.6.

Exercise 5.4 Let U be a subspace of a normed space V and let $A : U \to W$ be a linear operator to a normed space W. Suppose that $B : V \to W$ is a linear operator which extends A. Show that $\|A\| \leq \|B\|$.

Exercise 5.5 Show that point-wise and strong convergence in $\mathbf{B}(\mathbb{R}^n, \mathbb{R}^m)$ coincide.

Exercise 5.6 Prove that every infinite dimensional normed space V either over the field $K = \mathbb{R}$ or the field $K = \mathbb{C}$ admits a discontinuous linear operator $A : V \to K$. (Hint: Use a Hamel basis.)

Exercise 5.7 Show that Lemma 5.1 may fail if the domain of the linear operator is not a Banach space.

Exercise 5.8 Let $\{A_m\}_{m\geq 1}$ be a sequence of operators $A_m : V \to W$ between normed spaces. Prove that if $\{A_m\}_{m\geq 1}$ is Cauchy, then $\{\|A_m\|\}_{m\geq 1}$ is Cauchy. Does the converse implication hold as well?

Exercise 5.9 Let $A : V \to W$ be a linear operator between normed spaces. Prove that if A is open, then A is surjective.

Exercise 5.10 Give an example of a normed space V and a proper dense linear subspace of it.

5.2 Fixed-Point Techniques in Banach Spaces

Since a Banach space $\mathscr{B}$ is in particular a complete and non-empty metric space, the Banach Fixed-Point Theorem (Theorem 4.8) guarantees that any contraction on $\mathscr{B}$ admits a unique fixed-point. In more detail, a contraction on a Banach space $\mathscr{B}$ is a function (not necessarily linear) $F : \mathscr{B} \to \mathscr{B}$ such that there exists $0 < \alpha < 1$ with

$$\|Fx - Fy\| \leq \alpha \|x - y\|.$$

The guaranteed unique fixed-point for F is an element $x_0 \in \mathscr{B}$ with $F(x_0) = x_0$. From the proof of the Banach Fixed-Point Theorem, the fixed-point x_0 is obtained as the limit $x_k \to x_0$ where $x_{k+1} = F(x_k)$, and $x_1 \in \mathscr{B}$ is an arbitrary element.

It follows that any problem in a Banach space $\mathscr{B}$ whose solution can be expressed as a fixed-point for a contraction on $\mathscr{B}$ can be solved iteratively. We explore this technique in three different scenarios, namely for solving systems of linear equations, for solving first order differential equations, and for solving integral equations.

Remark 5.4 Notice that the more general problem

$$x_0 = F(x_0) + b$$

is subsumed by the technique described above since for $F_b(x) = F(x) + b$ one has that

$$\|F_b(x) - F_b(y)\| = \|F(x) - F(y)\|$$

and thus F is a contraction if, and only if, F_b is a contraction. Moreover, a fixed-point for F_b is precisely a solution of the equation $x_0 = F(x_0) + b$. It is customary to simplify the notation and write $x = Fx + b$. Notice that if F is a linear operator, then F is a contraction if, and only if, $\|F\| < 1$.

5.2.1 Systems of Linear Equations

Consider a linear system of n equations in n unknowns:

$$\sum_{i=1}^{n} a_{ki} x_i = b_k, \quad k = 1, 2, \ldots, n.$$

In terms of operators this system of equations can be written as $Ax = b$, where A is the corresponding $n \times n$ matrix of coefficients viewed as a linear operator $\mathbb{R}^n \to \mathbb{R}^n$, with $x = (x_1, \ldots, x_n)$ representing the unknown vector, and $b = (b_1, b_2, \ldots, b_n)$ the vector of free coefficients. Theoretically precise methods for solving such systems of linear equations are computationally demanding. For instance, Cramer's rule requires the computation of n^2 determinants, and thus, for large n, this becomes unfeasible. Moreover, numerical difficulties resulting from rounding may render some computation techniques unreliable. We now investigate the applicability of an iterative fixed-point technique in this case. Such a solution is often more robust than naive applications of theoretical exact solutions (though we will not delve into a numerical analysis of the solution here).

Since $Ax = b$ if, and only if, $x = (-A + I_n)x + b$, where I_n is the $n \times n$ identity matrix, we are led to consider the equation

$$x = Fx + b$$

where $F = -A + I_n$. The original problem is equivalent to the fixed-point problem

$$x = F_b x$$

where $F_b x = Fx + b = (-A + I_n)x + b$ is viewed as an operator $F_b : \mathbb{R}^n \to \mathbb{R}^n$. Let us assume some norm on $\mathbb{R}^n$ is chosen. For the Banach Fixed-Point Theorem to be applicable, F_b must be a contraction, and since F is a linear operator we must examine the norm $\| F \|$. In particular, if $\| F \| < 1$, then F, and thus F_b, is a contraction and the original problem is amenable to the iterative solution $x_{k+1} = F_b x_k$ with an arbitrary initial value $x_1 \in \mathbb{R}^n$.

Whether F is a contraction is typically strongly dependent on the chosen norm for the ambient space $\mathbb{R}^n$. We consider two cases below, recalling that the Kronecker δ function is given by

$$\delta_{k,i} = \begin{cases} 1 & \text{if } k = i, \\ 0 & \text{if } k \neq i. \end{cases}$$

Example 5.6 Consider $\mathbb{R}^n$ with the ℓ_∞ norm: $\|x\| = \max_{1 \le k \le n} |x_k|$. Then, for any $n \times n$ matrix C

$$\begin{aligned} \|Cx - Cy\| &= \max_{1 \le k \le n} \left| \sum_{i=1}^{n} c_{k,i}(x_i - y_i) \right| \le \max_{1 \le k \le n} \left[\sum_{i=1}^{n} |c_{k,i}| \, |x_i - y_i| \right] \\ &\le \max_{1 \le k \le n} \left[\sum_{i=1}^{n} |c_{k,i}| \right] \max_{1 \le i \le n} |x_i - x_i| = \max_{1 \le k \le n} \left[\sum_{i=1}^{n} |c_{k,i}| \right] \|x - y\|. \end{aligned}$$

Thus, if the inequality

$$\max_{1\le k\le n}\left[\sum_{i=1}^{n}|c_{k,i}|\right]<1$$

holds, then the operator C is a contraction. Returning to the system $Ax = b$ and the related operator $F_b x = (-A + I_n)x + b$, we see that F_b is guaranteed to be a contraction if

$$\max_{1\le k\le n}\sum_{i=1}^{n}|-a_{k,i}+\delta_{k,i}|<1,$$

in which case the original system can be solved iteratively.

Example 5.7 Let us now consider $\mathbb{R}^n$ with the ℓ_2 norm $\|x\|^2 = \sum_{k=1}^{n} x_k^2$, giving rise to the Euclidean metric. For an $n \times n$ matrix C viewed as an operator on $\mathbb{R}^n$, we have

$$\|Cx\|^2 = \sum_{k=1}^{n}(c_k \cdot x)^2$$

where c_k is the k-th row of the matrix C, and $c_k \cdot x$ is the standard inner product in $\mathbb{R}^n$. Applying the Cauchy-Schwarz Inequality we obtain

$$\|Cx\|^2 \le \sum_{k=1}^{n}\|c_k\|^2\|x\|^2$$

and thus we see that

$$\|C\| \le \sqrt{\sum_{k=1}^{n}\|c_k\|^2} = \sqrt{\sum_{k,i=1}^{n}c_{k,i}^2}.$$

Going back to the linear system $Ax = b$, we conclude that it is iteratively solvable if

$$\sum_{k,i=1}^{n}(-a_{k,i}+\delta_{k,i})^2 < 1.$$

5.2.2 Cauchy's Problem and the Volterra Equation

Consider Cauchy's problem for the first order differential equation:

$$\frac{dx}{dt}(t) = f[t, x(t)] \quad \text{with} \quad x(t_0) = x_0.$$

The associated integral equation

$$x(t) = \int_{t_0}^{t} ds\ f[s, x(s)] + x_0$$

is called a *Volterra integral equation*. Suppose that, for $|t - t_0| \leq \varepsilon$ and $|x - x_0| \leq \eta$, the function $f(t, x)$ is continuous and $|f(t, x)| < M$. Noting that a solution of the Volterra equation is also a solution to the original Cauchy problem, we introduce the operator

$$(Bx)(t) = \int_{t_0}^{t} ds\ f[s, x(s)] + x_0$$

since its fixed-points are precisely solutions to the Volterra equation. If B can be seen to be an operator

$$B : C(I, \mathbb{R}) \to C(I, \mathbb{R})$$

for a suitable interval $I \subseteq [a, b]$, and if B is moreover a contraction with respect to some norm, then locally Cauchy's problem has a unique solution, which, starting with an arbitrary x_0, may be computed iteratively:

$$x_1(t) = \int_{t_0}^{t} ds\ f(s, x_0) + x_0,\quad x_2(t) = \int_{t_0}^{t} ds\ f[s, x_1(s)] + x_0, \ldots,\ x_n(t) = \ldots$$

Since the norm we choose must turn the space $C(I, \mathbb{R})$ into a Banach space (since otherwise Banach's Fixed-Point Theorem does not apply) the only norm of those encountered so far that we can consider is the L_∞ norm (and now the importance of the L_p spaces becomes apparent as they are complete and thus they allow one to consider other norms).

Let us examine under which assumptions is it guaranteed that B is a contraction for a suitable interval. Consider $C(I, \mathbb{R})$ where $I = [t_0 - \varepsilon, t_0 + \varepsilon]$ and choose some interval $J = [x_0 - \eta, x_0 + \eta]$ contained in I. We seek to identify conditions on ε, η, and M that will assure the existence of a solution, at least locally. We note the following.

1. The function Bx is certainly a continuous function, given the assumptions on $f[t, s]$.
2.

$$\|Bx - x_0\| = \max_{t \in I} \left| \int_{t_0}^{t} ds\ f[s, x(s)] \right|$$
$$\leq \max_{t \in I} |t - t_0| \max_{t \in I,\ x \in J} |f(t, x)| \leq \varepsilon M.$$

We thus see that the operator B transforms the restriction of x to J into a function taking values in J provided that $\varepsilon \leq \eta/M$, a condition we now assume is met.

3. For all $x, y \in J$ we have that

$$\|Bx - By\| \leq \max_{t \in I} \int_{t_0}^{t} ds\, \big| f[s, x(s)] - f[s, y(s)] \big|.$$

4. Suppose the function $f(t, x)$ is not just continuous and bounded, but also satisfies the *Lipschitz* condition:

$$\big| f(t, x) - f(t, y) \big| \leq K|x - y|, \quad \forall\, x, y \in J,$$

uniformly for all $t \in I$. Then

$$\begin{aligned} \|Bx - By\| &\leq \max_{t \in I} \int_{t_0}^{t} ds\, \big| f[s, x(s)] - f[s, y(s)] \big| \leq K \max_{t \in I} \int_{t_0}^{t} ds\, \big| x(s) - y(s) \big| \\ &\leq K\varepsilon \max_{s \in I} \big| x(s) - y(s) \big| = K\varepsilon \|x - y\|, \quad \forall\, x, y \in J. \end{aligned}$$

Collecting these observations together we conclude that if $\varepsilon \leq \eta/M$ and $\varepsilon \leq 1/K$, then the operator B is a contraction and Cauchy's problem has one, and only one, solution inside the *reduced interval* $I = [t_0 - \varepsilon, t_0 + \varepsilon]$, with $\varepsilon \leq \min(\eta/M, 1/K)$. Furthermore, the solution can be obtained iteratively as a fixed-point solution of the associated Volterra equation.

5.2.3 Fredholm Equations

The Volterra equations, considered above, are examples of integral equations with variable limits. We now turn to consider *Fredholm equations*, which belong to the family of integral equations with fixed limits.

First consider the general form of a non-linear Fredhom equation

$$x(t) = \lambda \int_{a}^{b} ds\, K\big[t, s, x(s)\big] = [Ax](t),$$

where we consider A as an operator, and we seek to identify conditions under which the iterative fixed-point method is applicable. Let $K(t, s, x)$ be a continuous function in $[a, b] \times [a, b] \times [-\eta, \eta]$ and bounded by $M > 0$. Let $J = [-\eta, \eta]$ and consider

$C(J, \mathbb{R})$ with the L_∞ norm. Noticing that

$$\|Ax\| \leq \max_{t\in[a,b]} \left|\lambda \int_a^b ds\ K\big[t, s, x(s)\big]\right| \leq |\lambda| M(b-a)$$

we see that if $|\lambda| \leq \eta/\big[M(b-a)\big]$, then $\|Ax\| \leq \eta$ and so the operator A is well-defined when restricting to the interval J. Furthermore,

$$\begin{aligned}\|Ax - Ay\| &= \max_{t\in[a,b]} \left|\lambda \int_a^b ds\ \Big\{K\big[t, s, x(s)\big] - K\big[t, s, y(s)\big]\Big\}\right| \\ &\leq |\lambda| \max_{t\in[a,b]} \int_a^b ds\ \Big|K\big[t, s, x(s)\big] - K\big[t, s, y(s)\big]\Big|.\end{aligned}$$

If we also make the assumption that the function $K(t, s, x)$ is Lipschitz with respect to x, uniformly in the square $[a, b] \times [a, b]$, with Lipschitz constant K, then:

$$\|Ax - Ay\| \leq |\lambda|(b-a)K \max_{s\in[a,b]} |x(s) - y(s)| \leq |\lambda|(b-a)K\ \|x - y\|.$$

It thus follows that if $|\lambda|(b-a)K < 1$, then the operator A is a contraction, and thus the original non-linear Fredholm equation has a unique solution, obtained by the iterative scheme

$$x_{n+1}(t) = \int_a^b ds\ K\big[t, s, x_n(s)\big],$$

with $x_0(t)$ an arbitrary continuous function on $[a, b]$, satisfying $\max_t |x_0(t)| \leq \eta$. The condition

$$|\lambda|(b-a) \leq \min\{\eta/M, 1/K\}$$

guarantees the applicability of the iterative solution.

Consider now the linear Fredholm equation:

$$x(t) = \lambda \int_a^b ds\ K(t, s)x(s) + f(t),$$

with λ a real parameter, $f(t)$ a continuous function in $[a, b]$ and $K(t, s)$ continuous in the square $[a, b] \times [a, b]$ and bounded by $M > 0$. Still considering $C([a, b], \mathbb{R})$ with the L_∞ norm, consider the operator

$$(Bx)(t) = \lambda \int_a^b ds\ K(t,s)x(s) + f(t).$$

We now have that,

$$\|Bx - By\| = \max_{t\in[a,b]} \left|\lambda \int_a^b ds\ K(t,s)\big[\,x(s) - y(s)\big]\right|$$
$$\leq |\lambda| M(b-a) \max_{s\in[a,b]} \big|x(s) - y(s)\big| = |\lambda| M(b-a)\|x - y\|,$$

and thus, for $|\lambda| < 1/[M(b-a)]$, the operator B is a contraction and the uniform limit (i.e., with respect to the L_∞ norm) of the iterative process

$$x_{n+1}(t) = \lambda \int_a^b ds\ K(t,s)x_n(s) + f(t) = \dots$$

exists and gives the solution to the given Fredholm equation. A possible choice of the initial function is given by $x_0(t) = f(t)$.

Example 5.8 Consider in $C([0,1],\mathbb{R})$ the Fredholm equation:

$$x(t) = \lambda \int_0^1 ds\ tsx(s) + \alpha t,$$

with α a real parameter. Since $\max_{t,s\in[0,1]} K(t,s) = 1$ and $(b-a) = 1$, it follows that for $|\lambda| < 1/\big[M(b-a)\big] = 1$ the operator is a contraction and one can solve the problem by successive approximations:

$$x_0 = \alpha\, t = \alpha_0\, t,\;\; x_1 = \lambda \int_0^1 ds\ (t\, s)\ (\alpha_0\, s) + \alpha\, t = (\alpha + \alpha_0\, \lambda/3)\, t = \alpha_1\, t, \dots$$

$$x_{n+1} = \lambda \int_0^1 ds\ (t\, s)\ (\alpha_n\, s) = (\alpha + \alpha_n\, \lambda/3)\, t = \alpha_{n+1}\, t, \dots$$

For $|\lambda| < 1$ the limit $x = \lim_{n\to\infty} x_n$ exists and thus also $\beta = \lim_{n\to\infty} \alpha_n$ exists. When passing to the limit in the last equality we obtain that $\alpha + \beta\lambda/3 = \beta$, from which we conclude that $\beta = 3\alpha/(3-\lambda)$ and, finally, that $x(t) = 3\alpha/(3-\lambda)t$.

We got this solution by assuming $|\lambda| < 1$, but one can easily see that the limit of α_n as $n \to \infty$ also exists if $|\lambda| < 3$, as suggested by the last relation. In fact:

$$\alpha_{n+1} = \alpha + \alpha_n\, \lambda/3 = \cdots = \alpha \left[1 + \lambda/3 + (\lambda/3)^2 + \cdots + (\lambda/3)^{n+1}\right]$$

$$\Longrightarrow \quad \alpha_n(t) \to 3\alpha/(3-\lambda) \quad \text{for} \quad |\lambda| < 3.$$

The condition $|\lambda| < 1$ is a sufficient condition for the convergence of the iterative process which, as just seen, has a broader domain of convergence. More generally, the function $x(t)$ found above is actually a solution of the original problem for every $\lambda \neq 3$, as is easily checked with a test function $x(t) = \gamma t$, with γ unknown.

Exercises

Exercise 5.11 Show that a contraction on an arbitrary normed space need not have a fixed-point.

Exercise 5.12 Give an example of an isometry $F : \mathbb{R} \to \mathbb{R}$ without any fixed-points (here $\mathbb{R}$ is given the Euclidean metric structure).

Exercise 5.13 A function $f : X \to Y$ between metric spaces is a *Lipschitz* function if there exists a number $K \in \mathbb{R}$, called a *Lipschitz constant* such that

$$d(f(x), f(x')) \leq Kd(x, x')$$

for all $x, x' \in X$. Show that, with respect to the Euclidean metric on $\mathbb{R}$, a function $f : \mathbb{R} \to \mathbb{R}$ with a bounded derivative is Lipschitz.

Exercise 5.14 Prove that a Lipschitz function $f : X \to Y$ between metric spaces is uniformly continuous, but the converse may fail.

Exercise 5.15 Give an example of a function $F : X \to Y$ between metric spaces which satisfies $d(Fx, Fy) < d(x, y)$ for all $x, y \in X$, but is not a contraction.

Exercise 5.16 Perform a similar analysis as done in Example 5.7 but consider the ℓ_1 norm on $\mathbb{R}^n$ and attempt to obtain a reasonable criterion for the applicability of the fixed-point technique for solving a system of linear equations.

Exercise 5.17 Referring to Example 5.4, prove, once more, that the differentiation operator d/dt on $C([a, b], \mathbb{R})$ is unbounded by considering the sequence

$$x_n(t) = \sin[n\pi(b-t)/(b-a)].$$

Exercise 5.18 With successive substitutions, solve the Volterra equation:

$$x(t) = t - \int_0^t ds(t-s)\, x(s).$$

Exercise 5.19 Following Example 5.8, solve the equation:

$$x(t) = \frac{5}{6}t + \frac{1}{2}\int_0^1 ds\, ts\, x(s).$$

Exercise 5.20 Following Example 5.8, solve the equation:

$$x(t) = e^t - 1 + \int_0^1 ds\, t\, x(s).$$

5.3 Inverse Operators

From Chap. 2 we know that if a linear operator $A : V \to W$ admits an inverse function $A^{-1} : W \to V$, then that function is automatically a linear operator. The Open Mapping Theorem implies that if the linear spaces are Banach spaces and the operator is bounded, then its inverse is also bounded. In this section we will establish a condition on A that guarantees the existence of a bounded inverse without assuming the spaces are Banach. We will them obtain a simple yet powerful result which presents the inverse of an operator in terms of a convergent series of operators. The analysis will reveal the close connection between invertibility and fixed-point solutions. We will then have a second, deeper, look at the Volterra and Fredholm equations in light of the results below.

5.3.1 Existence of Bounded Inverses

We present conditions that guarantee the existence of inverse operators, providing estimates for their norm, and, under suitable conditions, we obtain the inverse as the sum of a strongly convergent series.

Theorem 5.13 *Let $A : V \to W$ be a bounded surjective linear operator between normed spaces. Then A is invertible with bounded inverse A^{-1} if, and only if, there exists a constant $C > 0$ such that $\|Ax\| \geq C\|x\|$, for all $x \in V$. Moreover, with such a constant C the operator norm of the inverse satisfies $\|A^{-1}\| \leq 1/C$.*

Proof Suppose that $\|Ax\| \geq C\|x\|$ holds. Then if $Ax = 0$, then $0 = \|Ax\| \geq C\|x\|$, and thus $x = 0$. In other words, $\text{Ker}(A) = \{0\}$, and thus A is injective. Since it is given that A is surjective, we conclude that A is invertible and that its inverse A^{-1} is automatically a linear operator. Moreover, for all $y \in W$ let $x \in V$ with $Ax = y$, and thus $x = A^{-1}y$. We then have

$$\|y\| = \|Ax\| \geq C\|x\| = C\|A^{-1}y\|,$$

showing that $\|A^{-1}\| \leq 1/C$. In the other direction, suppose that A^{-1} exists and is bounded. Then, for all $x \in V$

$$\|x\| = \|A^{-1}Ax\| \leq \|A^{-1}\|\|Ax\|.$$

Noticing that the norm of an invertible operator is strictly positive, we conclude that

$$\|Ax\| \geq \frac{1}{\|A^{-1}\|}\|x\|$$

and thus we may take $C = 1/\|A^{-1}\|$. □

Consider now an in-homogenous fixed-point problem $x = Ax + b$, where $A : \mathscr{B} \to \mathscr{B}$ is a linear operator on a Banach space $\mathscr{B}$. As we know, if A is a contraction, namely if $\|A\| < 1$, then a unique solution exists. Notice though that the problem may be re-written as $(I - A)x = b$, where we write I for the identity operator on $\mathscr{B}$. Now, if $(I - A)$ is invertible, then we obtain the unique solution $x = (I - A)^{-1}b$. Since the latter does not rely on the norm it actually holds in any linear space. We thus have two criteria for the solution of the equation $x = Ax + b$, one is that A be a contraction and the other is that $I - A$ be invertible. The next result shows that the former implies the latter, when the ambient space is a Banach space.

Proposition 5.3 *Let* $A : \mathscr{B} \to \mathscr{B}$ *be a linear operator on a Banach space* $\mathscr{B}$*. If* $\|A\| < 1$*, then* $I - A$ *is invertible. Moreover,* $\|(I - A)^{-1}\| \leq 1/(1 - \|A\|)$.

Proof Since $\|Ax\| \leq \|A\|\|x\|$, if follows that

$$\|(I - A)x\| = \|x - Ax\| \geq \|x\| - \|Ax\| \geq \|x\| - \|A\|\|x\| = (1 - \|A\|)\|x\|,$$

and the claim would follow from Theorem 5.13 if we can show that $I - A$ is surjective. In other words, given $y \in W$, we wish to solve the equation

$$(I - A)x = y,$$

or, equivalently, the equation

$$x = Ax + y.$$

But this is precisely an in-homogenous fixed-point problem, and since A is assumed to be a contraction, a solution exists. □

We thus see that the invertibility of $I - A$ is important for the solution of fixed-point problems, and thus, as we saw above, to the solution of systems of linear equations, differential equations, and integral equations (and other problems as well which we did not mention). While the condition $\|A\| < 1$ gives a sufficient condition for the existence of $(I - A)^{-1}$, it does not provide us with any feasible means of obtaining the inverse. The following result remedies the situation.

Definition 5.9 Let $A : V \to V$ be a linear operator on a normed space V. The series $\sum_{k=0}^{\infty} A^k$, where A^0 is interpreted as the identity operator, is called the *Neumann series* of A.

Recall the notation $\mathbf{B}(V, W)$ for the set of all bounded linear operators $A : V \to W$. When $V = W$ we will write $\mathbf{B}(V)$ instead of $\mathbf{B}(V, W)$. Similarly, for a Banach space $\mathscr{B}$, we write $\mathbf{B}(\mathscr{B})$ instead of $\mathbf{B}(\mathscr{B}, \mathscr{B})$.

Theorem 5.14 *Let $A : V \to V$ be a linear operator on a normed space V. If the Neumann series for A converges in the operator norm, then the Neumann series is the inverse of $I - A$.*

Proof Suppose that the Neumann series converges to the operator S, that is

$$S = \lim_{m \to \infty} \sum_{k=0}^{m} A^k$$

with respect to the operator norm in the space $\mathbf{B}(V)$. Now, since

$$(I - A) \sum_{k=0}^{m} A^k = I - A^{m+1},$$

when m tends to ∞ we obtain

$$(I - A) \sum_{k=0}^{\infty} A^k = I - \lim_{m \to \infty} A^{m+1}.$$

We leave it as an exercise to the reader to verify that if the Neumann series of A converges, then $A^m \to 0$ in the operator norm, and thus we see that the Neumann series is a right inverse of $I - A$. A similar agument shows it is also a left inverse, and the proof is complete. □

The expected sufficient condition on an operator A on a Banach space guaranteeing the convergence of the Neumann series is given in the next result.

Corollary 5.3 *Let $A : \mathscr{B} \to \mathscr{B}$ be a linear operator on a Banach space $\mathscr{B}$ with $\|A\| < 1$. Then the Neumann series of A converges, and in particular*

$$(I - A)^{-1} = \sum_{k=0}^{\infty} A^k,$$

where the series converges in the operator norm.

Proof All we have to do is establish the convergence in $\mathbf{B}(\mathscr{B})$ of the Neumann series. Since $\mathscr{B}$ is a Banach space, so is the space $\mathbf{B}(\mathscr{B})$. Using Theorem 5.6, it suffices to show the convergence of the series $\sum_{k=0}^{\infty} \|A^k\|$. But $\|A^k\| \leq \|A\|^k$ and thus the latter series is majorized by $\sum_{k=0}^{\infty} \|A\|^k$, which converges since $0 \leq \|A\| < 1$. □

Corollary 5.4 *If A is as above and $\lambda \in \mathbb{R}$ is such that $\|\lambda A\| = |\lambda|\|A\| < 1$, then $I - \lambda A$ is invertible and its inverse is given by the Neumann series*

$$(I - \lambda A)^{-1} = \sum_{k=0}^{\infty} \lambda^k A^k.$$

5.3.2 Fixed-Point Techniques Revisited

The fixed-point techniques given in Sect. 5.2 relied on point-wise convergence. The convergence of the Neumann series though is strong convergence (i.e., in the operator norm), a form of convergence much stronger than point-wise convergence. We thus expect to have greater control over the situation when the Neumann series can be brought into the playground when solving a fixed-point problem. We now look again at some of the differential and integral equations we solved above and witness the enhancement due to the strong convergence.

Example 5.9 Consider the linear Fredholm equation

$$x(t) = \lambda \int_a^b ds\ K(t,s)x(s) + f(t) = \lambda Ax + f$$

with $K(t, s)$, known as the *nucleus*, continuous in the square $[a, b] \times [a, b]$, and $f(t)$ continuous in the interval $[a, b]$. Since

$$|K(t,s)| \leq M \quad \Longrightarrow \quad \|A\| \leq |\lambda| M(b-a) = q$$

we see that for those values of λ such that $q < 1$, the Neumann series

$$\sum_{k=0}^{\infty} A^k$$

of A converges (in the operator norm). Here $A^0 = \text{id} = I$ and

$$A^1 = \int_a^b ds\, K_1(t,s)(*),\quad A^2 = \int_a^b ds\, K_2(t,s)(*), \ldots,$$

$$A^n = \int_a^b ds\, K_n(t,s)(*), \ldots,$$

where

$$K_1(t,s) = K(t,s),\quad K_2(t,s) = \int_a^b du\, K(t,u)\, K(u,s)\,, \ldots,$$

$$K_n(t,s) = \int_a^b du\, K_{n-1}(t,u)\, K(u,s), \ldots$$

Notice further that for all $t, s \in [a, b]$

$$|\lambda^i K_i(t,s)| \leq |\lambda|^i M^i (b-a)^i \leq q^i,$$

and, since $0 \leq q < 1$ by assumption, the series

$$K(t,s;\lambda) = \sum_{i=1}^{\infty} \lambda^i K_i(t,s),$$

converges uniformly in $[a, b] \times [a, b]$. The function $K(t, s; \lambda)$ is called the *resolvent nucleus* and is a continuous function in the square $[a, b] \times [a, b]$ (as a uniform limit of continuous functions). From here it follows that

$$x(t) = f(t) + \int_a^b ds\, K(t,s;\lambda) f(s).$$

Example 5.10 Consider the Fredholm integral equation

$$x(t) = \sin t - \frac{t}{4} + \frac{1}{4} \int_0^{\pi/2} ds\, ts\, x(s).$$

With reference to the previous example, we have the estimates

$$|K(t,s)| = |ts| \leq \frac{\pi^2}{4},\quad (b-a) = \frac{\pi}{2},\quad \lambda = \frac{1}{4} < \frac{1}{M(b-a)} = \frac{8}{\pi^3}$$

and thus we can perform the same series expansion as above. The resolvent nucleus is now given by

$$K(t, s; \lambda) = \lambda K_1(t, s) + \lambda^2 K_2(t, s) + \lambda^3 K_3(t, s) + \cdots,$$

where

$$K_1(t, s) = t\,s, \quad K_2(t, s) = \int_0^{\pi/2} du\,(t\,u)\,(u\,s) = \alpha\,t\,s,$$

$$K_3(t, s) = \int_0^{\pi/2} du\; K_2(t, u) K_1(u, s) = \alpha \int_0^{\pi/2} du\,(t\,u)\,(u\,s) = \alpha^2 t\,s, \ldots$$

$$K_n(t, s) = \alpha^{n-1} t\,s + \cdots \quad \text{with} \quad \alpha = \int_0^{\pi/2} du\,u^2 = \pi^3/24.$$

The expansion of all nuclei is given by:

$$K(t, s; \lambda) = \lambda\left[t\,s + \lambda\alpha\,t\,s + (\lambda\alpha)^2 t\,s + \cdots\right] = \lambda\,t\,s\,(1 - \lambda\alpha)^{-1},$$

since $\lambda\alpha = \pi^3/(4.24) < 1$. The solution of our problem is then

$$x(t) = \sin t - \frac{t}{4} + \frac{1}{4}\int_0^{\pi/2} ds\,\frac{1}{1 - \alpha/4}\,t\,s\left(\sin s - \frac{s}{4}\right) =$$

$$\sin t - \frac{t}{4} + \frac{t}{4 - \alpha}\int_0^{\pi/2} ds\left(s \sin s - \frac{s^2}{4}\right).$$

Since

$$\int_0^{\pi/2} s \sin s = -s \cos s\,\Big|_0^{\pi/2} + \int_0^{\pi/2} ds\,\cos s = \sin s\,\Big|_0^{\pi/2} = 1$$

we finally obtain

$$x(t) = \sin t - \frac{t}{4} + \frac{t}{4 - \alpha}\left(1 - \frac{\alpha}{4}\right) = \sin t - \frac{t}{4} + \frac{t}{4} = \sin t.$$

Let us now reconsider the Volterra equation, in light of the more powerful technique of the Neumann series.

Example 5.11 Consider in $C([a, b], \mathbb{R})$ the linear Volterra equation

$$x(t) = \lambda \int_a^t ds\ K(t, s)\, x(s) + f(t) = \lambda Ax + f,$$

with the usual assumption of continuity and boundedness on $K(t, s)$ and $f(t)$, and let

$$M = \max_{\{t,s\}\in[a,b]} |K(t, s)|, \quad F = \max_{t\in[a,b]} |f(t)|.$$

Compared with the previous treatment of the Fredholm equation by means of pointwise fixed-point solutions, here the integral does not extend over the entire interval $[a, b]$ but only up to t, which in turn becomes the integration variable of the next iteration. Since usually integration makes functions *smoother*, we expect to obtain better conditions on the values of λ for which the iterative solution is converging. Then, consider the solution

$$x(t) = \sum_{i=0}^{\infty} \lambda^i A^i f = \sum_{i=0}^{\infty} \lambda^i \phi_i,$$

which certainly exists at least for those values of λ found for the linear Fredholm equation. The following estimates hold

$$|\phi_1(t)| \le \int_a^t ds\ |K(t, s)\, f(s)| \le M\ F\ (t - a),$$

$$|\phi_2(t)| \le \int_a^t ds\ |K(t, s)\, \phi_1(s)| \le M^2\ F \int_a^t ds\ (s - a) = M^2 F(t - a)^2/2, \ldots,$$

$$|\phi_n| \le M^n F\ \frac{(t - a)^n}{n!}, \ldots$$

It follows that

$$\left|\sum_{i=0}^{\infty} \lambda^i \phi_i\right| \le \sum_{i=0}^{\infty} |\lambda^i|\, |\phi_i| \le F \sum_{i=0}^{\infty} |\lambda|^i M^i\ \frac{(t - a)^i}{i!} = F \exp\big[\, |\lambda|\, M\, (t - a)\big],$$

is convergent for all values of λ. Therefore, the function $x(t)$ is continuous and is a solution of the Volterra equation for every λ.

Example 5.12 Let us see how the linear homogeneous Fredholm equation admits solutions other than the zero solution. Given the equation

$$x(t) = \lambda \int_0^1 ds\ (t+s)\ x(s),$$

we introduce the constants c_1 and c_2 given by

$$x(t) = \lambda t \int_0^1 ds\ x(s) + \lambda \int_0^1 ds\ sx(s) = \lambda\, t\, c_1 + \lambda\, c_2.$$

By inserting this form of $x(t)$ into the first equation, one obtains

$$\lambda\, t\, c_1 + \lambda c_2 = \lambda \int_0^1 ds\ (t+s)(\lambda\, s\, c_1 + \lambda c_2) = \lambda^2 \Big(\frac{c_1}{2}\, t + c_2\, t + \frac{c_1}{3} + \frac{c_2}{2}\Big).$$

The coefficients of similar terms in t must be equal, so that

$$\begin{cases} c_1 = \lambda\, c_1/2 + \lambda\, c_2 \\ c_2 = \lambda\, c_1/3 + \lambda\, c_2/2. \end{cases}$$

The two equations are compatible for $\lambda = 6 \pm 4\sqrt{3}$ and the system, since it is homogeneous, admits infinitely many solutions, namely

$$c_1 = \frac{2\lambda}{2-\lambda}\, c, \quad c_2 = c,$$

with c an arbitrary parameter. The non-trivial solution of our equation is then given by

$$x(t) = \lambda \Big(\frac{2\lambda t}{2-\lambda} + 1\Big),$$

up to an arbitrary proportionality factor. Obviously, $|\lambda| = \big|6 \pm 4\sqrt{3}\big| > 1/2$, below which value the solution is unique.

Remark 5.5 In the previous examples the L_∞ norm has always been assumed. This was done in order to allow for the machinery developed thus far. It should be noted though that this does not exclude the existence of non-continuous solutions, i.e., ones that we have excluded a priori.

Example 5.13 Consider the homogeneous Volterra equation

$$x(t) = \int_0^t ds\, s^{t-s}\, x(s), \quad 0 \le t \le 1.$$

The nucleus $K(t, s) = s^{t-s}$ is continuous in the integration domain with $s \le t$, so that the equation admits the unique continuous solution $x = 0$. However, there exist infinitely many solutions not belonging to $C([a, b], \mathbb{R})$, given by $x(t) = c\, t^{t-1}$ with c an arbitrary constant, as can be checked directly

$$\int_0^t ds\, s^{t-s}\, x(s) = \int_0^t ds\, s^{t-s} s^{s-1} = \int_0^t ds\, s^{t-1} = \frac{s^t}{t}\Big|_0^t = t^{t-1} = x(t).$$

As expected, the function $x(t)$ is singular, since $x(t) \to \infty$ as $t \to 0$.

Exercises

Exercise 5.21 Prove that if the Neumann series of an operator $A : V \to V$, where V is a normed space, converges, then $A^m \to 0$ in the operator norm.

Exercise 5.22 If $A : V \to W$ is an injective linear operator between normed spaces, does it follow that $\|A\| > 0$?

Exercise 5.23 Show that linear operators can be as wild as one desires by showing that given any function $f : X \to Y$, where X is an arbitrary set and Y is a linear space over K, there exists a linear space $\hat{X}$ and an injection $i : X \to \hat{X}$ together with a linear operator $F : \hat{X} \to Y$ such that $f = F \circ i$.

Exercise 5.24 Consider the space $\mathbb{R}$ with its standard inner product structure and induced norm. Given any $a \in \mathbb{R}$ show that the function $\psi : \mathbb{R} \to \mathbb{R}$ given by

$$\psi(x) = ax$$

is a bounded linear operator. What is its norm? How many such operators have norm equal to 1?

Exercise 5.25 Repeat the previous exercise with $\mathbb{C}$ instead of $\mathbb{R}$.

Exercise 5.26 Let $\mathscr{B}$ be a Banach space and I the identity operator on $\mathscr{B}$. Prove that any linear operator $A : \mathscr{B} \to \mathscr{B}$ satisfying $\|A - I\| < 1$ is invertible.

Exercise 5.27 Let $\mathscr{B}$ be a Banach space and I the identity operator on $\mathscr{B}$. Suppose $A : \mathscr{B} \to \mathscr{B}$ is a linear operator with $\|A - I\| = 1$, is A necessarily not invertible?

Exercise 5.28 Let $B : \mathscr{B} \to \mathscr{B}$ be an invertible linear operator on a Banach space $\mathscr{B}$. Prove that any linear operator $A : \mathscr{B} \to \mathscr{B}$ satisfying $\|A - B\| < \|B\|$ is invertible.

Exercise 5.29 Let $B : \mathscr{B} \to \mathscr{B}$ be an invertible linear operator on a Banach space $\mathscr{B}$. If $A : \mathscr{B} \to \mathscr{B}$ is a linear operator satisfying $\|A - B\| > \|B\|$, is A necessarily not invertible?

Exercise 5.30 Let $A : \mathscr{B} \to \mathscr{B}$ be an invertible linear operator on a Banach space $\mathscr{B}$. Prove that given $\varepsilon > 0$ there exists a $\delta > 0$ such that $\|B^{-1} - A^{-1}\| < \varepsilon$ for all linear operators $B : \mathscr{B} \to \mathscr{B}$ satisfying $\|B - A\| < \delta$.

5.4 Dual Spaces

Let us briefly recall the fundamentals of duality for finite dimensional linear spaces, where for simplicity we take the ground field to be $\mathbb{R}$. The *dual space* of a finite dimensional linear space V is the linear space $V^* = \text{Hom}(V, \mathbb{R})$ of all linear operators $A : V \to \mathbb{R}$, where $\mathbb{R}$ is viewed as a linear space over itself in the usual way. Given a basis for V, one can associate with it its dual basis, which is then a basis for V^*, in particular showing that V and V^* have the same dimension, and thus they are isomorphic.

When considering these results in the infinite dimensional setting there are two immediate issues. The first is in the choice of linear operators $V \to \mathbb{R}$, namely do we consider all linear operators or only the bounded ones. The second issue is that results about the dual space that depend on a basis, for most infinite dimensional linear spaces, are only theoretical (since bases, while they exist, are often impossible to exhibit explicitly).

With that in mind, we define the dual space in the infinite dimensional context and we prove three major results. First is the Riesz Representation Theorem which gives a complete characterization of the dual space of any Hilbert space. Then we compute some duals related to the ℓ_p spaces, and finally we have a look at the Hahn-Banach Theorem (necessarily only a brief look, since a full treatment can easily fill an entire chapter).

5.4.1 Linear Functionals and the Riesz Representation Theorem

We now define linear functionals and the dual space of a general linear space, study several examples, and then establish the most general form of a bounded linear functional on a Hilbert space.

Definition 5.10 A *linear functional* on a normed space V is a linear operator $V \to K$ to the ground field K, thus either $K = \mathbb{R}$ or $K = \mathbb{C}$. The set of all bounded linear functionals $F : V \to K$ is denoted by V^* and is called the *dual space* of V.

Of course, the general theory of linear operators specializes to linear functionals. In particular,

1. The norm of a linear functional $F \in V^*$ is given by
$$\|F\| = \sup_{\|x\| \leq 1} |F(x)|$$
(or any of the other equivalent expressions given in Proposition 5.2).
2. A linear functional is continuous if, and only if, it is bounded.
3. V^* is a Banach space.

Example 5.14 Consider $C([a,b],\mathbb{R})$ with the L_∞ norm, i.e., $\|x\| = \max_{t\in[a,b]}\{|x(t)|\}$. The definite integral

$$F(x) = \int_a^b dt\, x(t)$$

defines a linear functional on V. Indeed, linearity follows by fundamental properties of integration, and as for the norm of F, the inequality

$$|F(x)| \leq \int_a^b dt\, |x(t)| \leq (b-a) \max_{a\leq t\leq b} |x(t)| = (b-a)\|x\|$$

shows that $\|F\| \leq b-a$. In fact, consideration of the constant function $x(t) = 1$, shows that $\|F\| = b-a$.

Example 5.15 The previous example easily generalizes as follows. Still considering $C([a,b],\mathbb{R})$ with the L_∞ norm, suppose a function $y(t)$ is given which is integrable over $[a,b]$. Then, integrating against y gives rise to the function $F_y\colon C([a,b],\mathbb{R}) \to \mathbb{R}$ given by

$$F_y(x) = \int_a^b dt\, y(t)\, x(t)$$

(the previous example is the case $y(t) = 1$). The verification that F_y is linear follows immediately and the computation showing that

$$\|F\| = \int_a^b dt\, |y(t)|$$

is left for the reader.

For a real Hilbert space $\mathscr{H}$ it is a non-trivial fact that the dual space $\mathscr{H}^*$ and the original space $\mathscr{H}$ are essentially the same, namely $\mathscr{H}$ is *self-dual*. Before we can establish this result we need to first look at linear functionals on inner product spaces and study perpendicularity in Hilbert spaces, an important and non-trivial issue.

Proposition 5.4 *Let V be an inner product space over the field K and let $y \in V$. The function $F_y : V \to K$ given by $F_y(x) = \langle y, x \rangle$ is a bounded linear functional with $\|F_y\| = \|y\|$. In particular, $F_y \in V^*$.*

Proof The linearity of F_y is immediate from the definition of an inner product space. Using the Cauchy-Schwarz Inequality, it follows that

$$|F_y(x)| = |\langle y, x \rangle| \leq \|y\| \|x\|$$

which thus shows that $\|F_y\| \leq \|y\|$. Considering the case where $x = y$ shows that $\|F_y\| = \|y\|$. □

A linear functional of the form F_y is said to be *representable*, with y *representing* it. For every inner product space V, we thus obtain a natural candidate for a comparison between V and its dual V^*, namely the function $\phi : V \to V^*$ given by $\phi(y) = F_y$, i.e., mapping every element $y \in V$ to the bounded linear functional it represents. This function is (easily seen to be) an injection, and so V always embeds in its dual space. However, ϕ need not be surjective, i.e., not every bounded linear functional on V need be representable. In this and in similar situations, a result establishing conditions under which ϕ is surjective are thus called *representability theorems*. The one we present below, namely that if V is a Hilbert space, then ϕ is surjective, is one of many famous representability theorems due to Riesz. We first need to establish some general geometric results in Hilbert spaces.

Theorem 5.15 *Any non-empty subset $S \subseteq \mathscr{H}$ of a Hilbert space $\mathscr{H}$ which is closed and satisfies $(x + y)/2 \in S$ for all $x, y \in S$, contains an element of smallest norm.*

Proof Consider the real number $\delta = \inf\{\|x\| \mid x \in S\}$, which exists since $S \neq \emptyset$. We need to show that the infimum is attained. There certainly exists a sequence $\{x_k\}_{k\geq 1}$ in S with $\|x_k\| \to \delta$. It suffices to show that $x_k \to x_0$ for some $x_0 \in \mathscr{H}$. Indeed, the continuity of the norm would then imply that $\|x_0\| = \delta$ and since S is closed, the limit point x_0 must itself be in S. Since $\mathscr{H}$ is complete the convergence of $\{x_k\}_{k\geq 1}$ would follow if we can show that the sequence is Cauchy. To that end, recall the parallelogram law

$$\|x + y\|^2 + \|x - y\|^2 = 2\|x\|^2 + 2\|y\|^2$$

which holds in any inner product space. Let now $x, y \in S$ and apply the parallelogram law to $x/2$ and $y/2$, to obtain

$$\frac{1}{4}\|x - y\|^2 = \frac{1}{2}\|x\|^2 + \frac{1}{2}\|y\|^2 - \|\frac{x + y}{2}\|^2.$$

Since $(x + y)/2 \in S$ by assumption, it follows that

$$\|x - y\|^2 \leq 2\|x\|^2 + 2\|y\|^2 - 4\delta^2.$$

Applying this inequality to the elements of the sequence $\{x_k\}_{k\geq 1}$ shows that it is indeed a Cauchy sequence, as required. □

For the next result recall that in an inner product space V the notation $x \perp y$ stands for $\langle x, y\rangle = 0$, and $x \perp S$, for a subset $S \subseteq V$, stands for $\langle x, y\rangle = 0$ for all $y \in S$. The geometric interpretation of $x \perp y$ is that x and y are perpendicular.

Theorem 5.16 *Let $\mathscr{H}$ be a Hilbert space and $S \subset \mathscr{H}$ a proper closed linear subspace of $\mathscr{H}$. There exists then an element $z \neq 0$ such that $z \perp S$.*

Proof Since S is a proper subset of $\mathscr{H}$, let $x \in \mathscr{H} - S$. Consider the set

$$x + S = \{x + y \mid y \in S\}$$

and notice that it satisfies the conditions of Theorem 5.15. Thus let z be an element of minimal norm in the set $x + S$. Note that $x \notin S$ implies $z \neq 0$, and we claim that $z \perp S$. Indeed, let $s \in S$ be arbitrary, where we may assume $\|s\| = 1$. Noting that $z - \alpha s \in x + S$ for all α in the ground field (either $\mathbb{R}$ or $\mathbb{C}$), the minimality of the norm of z implies that

$$\|z\|^2 \leq \|z - \alpha s\|^2 = \langle z - \alpha s, z - \alpha s\rangle = \|z\|^2 + |\alpha|^2\|s\|^2 - \alpha\langle z, s\rangle - \overline{\alpha}\langle s, z\rangle$$

which simplifies to

$$0 \leq |\alpha|^2 - \alpha\langle z, s\rangle - \overline{\alpha}\langle s, z\rangle.$$

Taking $\alpha = \langle s, z\rangle$ leads to

$$0 \leq |\alpha|^2 - \alpha\overline{\alpha} - \overline{\alpha}\alpha = -\|\alpha\|^2 = -\langle s, z\rangle^2$$

and the result follows. □

We can now prove the Riesz representation theorem.

Theorem 5.17 (Riesz Representation Theorem For Hilbert Spaces) *For a Hilbert space $\mathscr{H}$ over the field K the function $\phi : \mathscr{H} \to \mathscr{H}^*$, given by $\phi(y) = F_y$, where $F_y(x) = \langle y, x\rangle$, is an isometric (anti) isomorphism. In more detail:*

1. *ϕ is a bijection.*
2. *ϕ is norm preserving, i.e., $\|\phi(y)\| = \|y\|$ for all $y \in \mathscr{H}$.*
3. *ϕ is additive, i.e., $\phi(y + y') = \phi(y) + \phi(y')$ for all $x, y \in \mathscr{H}$.*
4. *If $K = \mathbb{R}$, then $\phi(\alpha y) = \alpha\phi(y)$.*
5. *If $K = \mathbb{C}$, then $\phi(\alpha y) = \overline{\alpha}\phi(y)$.*

Proof The validity of the last three claims is immediate from basic properties of the inner product. The fact that ϕ is norm preserving was already established in Proposition 5.4, and in particular ϕ is injective. It thus only remains to show that ϕ is surjective, in other words that every element in the dual space $\mathscr{H}^*$ is of the form F_y for some $y \in \mathscr{H}$ (i.e., is representable). This is the only part of the theorem where the completeness of $\mathscr{H}$ plays a role.

Let $F \in \mathscr{H}^*$ be an arbitrary bounded linear functional on $\mathscr{H}$. If $F = 0$ the claim is evident so we may proceed under the assumption that $F \neq 0$. Consider the kernel $S = \ker(F) = \{x \in \mathscr{H} \mid Fx = 0\} = F^{-1}(\{0\})$. Then S is a linear subspace of $\mathscr{H}$, and in fact a proper one by the assumption that $F \neq 0$. Further, since F is continuous, S is also closed (as the inverse image of the closed set $\{0\}$). By Theorem 5.16 there exists a non-zero $y \in \mathscr{H}$ such that $y \perp S$. In particular, $F(y) \neq 0$ and let $z = y/F(y)$, and note that $F(z) = 1$. Now, for all $x \in \mathscr{H}$

$$F(x - F(x)z) = F(x) - F(x)F(z) = 0$$

and thus $x - F(x)z \in S$. Since z is perpendicular to S it follows that

$$0 = \langle z, x - F(x)z\rangle = \langle z, x\rangle - F(x)\|z\|^2.$$

Let $h = z/\|z\|^2$ and then it follows that

$$F(x) = \langle h, x\rangle,$$

as required. □

5.4.2 Duals of Classical Spaces

For a real Hilbert space $\mathscr{H}$ the Riesz Representation Theorem establishes that $\mathscr{H}$ is isomorphic to its dual $\mathscr{H}^*$, and moreover the isomorphism preserves the norm. This prompts the following definition.

Definition 5.11 A function $A : V \to W$ between two normed spaces is a *linear isometry* if A is a linear isomorphism which respects the norms in the sense that $\|Ax\| = \|x\|$ for all $x \in V$. Normed spaces V and W are said to be *congruent* if there exists a linear isometry between them.

One can easily verify that a function $A : V \to W$ is a linear isometry if, and only if, it is a linear isomorphism which is also an isometry with respect to the induced metrics on V and W.

The following theorem establishes classical congruences for the family of ℓ_p spaces.

Theorem 5.18 (Riesz) *For the claims below all linear spaces are of real sequences, and thus are linear spaces over the field* $\mathbb{R}$.

1. $c_0^* \cong \ell_1$.
2. $\ell_1^* \cong \ell_\infty$.
3. $\ell_p^* \cong \ell_q$ *for all* $1 < p, q < \infty$ *such that* $1/p + 1/q = 1$.

Proof

1. Firstly, we construct a function $A : c_0^* \to \ell_1$, for which we use the *standard unit vectors* $e_k = (0, \ldots, 0, 1, 0, \ldots) \in c_0$, with 1 in the k-th position. Notice that $\{e_k\}_{k \geq 1}$ is not a basis for c_0. Given a linear functional $F \in c_0^*$ let AF be the sequence $(Fe_1, Fe_2, \ldots, Fe_k, \ldots)$, and denote $a_k = Fe_k$. Firstly, we claim that $AF \in \ell_1$, so that the codomain of A is indeed ℓ_1. To show that $AF \in \ell_1$ we need to establish that

$$\sum_{k=1}^{\infty} |a_k| < \infty.$$

For any $a \in \mathbb{R}$, let $\sigma(a)$ denote the sign of a, i.e., $\sigma(a) = a/|a|$ (with $\sigma(0) = 0$) and consider

$$x^{(m)} = (\sigma(a_1), \ldots, \sigma(a_m), 0, 0, 0, \ldots).$$

Since $|F(x^{(m)})| \leq \|F\| \|x^{(m)}\| = \|F\| < \infty$, and noting that

$$F(x^{(m)}) = \sum_{k=1}^{m} |a_k|,$$

it follows that

$$\sum_{k=1}^{m} |a_k| \leq \|F\| < \infty$$

and the same inequality remains valid when passing to the limit as $m \to \infty$. Having defined $A : c_0^* \to \ell_1$, the verification that A is linear is immediate, and thus it remains to show that A is surjective and that $\|AF\|_1 = \|F\|$. Indeed, for surjectivity let $a \in \ell_1$, i.e.,

$$\sum_{k=1}^{\infty} |a_k| < \infty.$$

If we now define $H : c_0 \to \mathbb{R}$ by means of the formula

$$Hx = \sum_{k=1}^{\infty} a_k x_k,$$

then, formally,

$$Fe_k = a_k$$

and thus

$$AH = a.$$

Thus, it remains to show that Hx is well-defined. Indeed,

$$\sum_{k=1}^{\infty} |a_k x_k| \le \sup |x_k| \sum_{k=1}^{\infty} |a_k| \le \|x\|_\infty \|a\|_1$$

and so the series defining Hx is absolutely convergent, and thus convergent. The linearity of H is obvious and the fact that it is bounded follows from the estimate above. It thus follows that indeed $H \in c_0^*$, and, as noted, $AH = a$, so that A is surjective. Combining the estimates above it also follows that $\|F\| = \|AF\|_1$, so A is an isometry.

2. The details of the proof are similar to the proof of the first claim, so we leave this proof to the reader.
3. The details are again similar to those given above, but this time some adaptation is required in the construction, with some complicating consequences in the needed estimates. Given $F \in \ell_p^*$ let $AF = (Fe_1, \ldots, Fe_k)$ where $\{e_k\}_{k\ge 1}$ are again the standard unit vectors, and denote again $a_k = Fe_k$. Considering

$$x^{(m)} = (|a_1|^{q-1}\sigma(a_1), \ldots, |a_m|^{q-1}\sigma(a_m), 0, 0, 0 \ldots)$$

we obtain

$$\sum_{k=1}^{m} |a_k|^q = F(x^{(m)}) \le \|F\|\|x^{(m)}\|_p = \|F\| \left(\sum_{k=1}^{m} (|a_k|^{q-1})^p\right)^{1/p} = \|F\| \left(|a_k|^q\right)^{1/p}.$$

In other words

$$\left(\sum_{k=1}^{m} |a_k|^q\right)^{1/p} \le \|F\|$$

and thus $\|a\|_q \le \|F\| < \infty$. This part of the proof then guarantees that $A : \ell_p^* \to \ell_q$ has the claimed codomain. The rest of the proof proceeds as the one above, namely defining for a given $a \in \ell_q$ the function $H : \ell_p \to \mathbb{R}$ by means of the formula

$$H(x) = \sum_{k=1}^{\infty} a_k x_k,$$

where Hölder's Inequality is used for the relevant estimates. □

Remark 5.6 Similar congruences hold true for L_p spaces. However, the proofs rely on non-trivial measure theoretic results, and thus lie beyond the scope of this book.

5.4.3 The Hahn-Banach Theorem

The main result we now present is a general result about normed spaces, valid not only for Banach spaces. We will present some of its consequences that further relate a space and its dual.

Theorem 5.19 (Hahn-Banach) *Let V be a normed space over K, with either $K = \mathbb{R}$ or $K = \mathbb{C}$. Suppose that $U \subseteq V$ a linear subspace, and $f : U \to K$ a bounded linear functional on the subspace U. There exists then a linear functional $F : V \to K$ such that $F|_U = f$ and such that $\|F\| = \|f\|$.*

Proof For simplicity, we will only give the proof for real normed spaces. Thus we assume that a bounded linear functional $f : U \to \mathbb{R}$ is given. Clearly, if $f = 0$, then $F = 0$ is the desired extension to all of V. We may thus proceed under the assumption that $f \neq 0$, and thus, normalizing if needed, that $\|f\| = 1$. Suppose that $x_0 \in V - U$ and let U_1 be the span of $U \cup \{x_0\}$. It is easy to see that U_1 consists of all vectors of the form $x + \beta x_0$, where $x \in U$ and $\beta \in \mathbb{R}$. Notice that, given any $\alpha \in \mathbb{R}$, the function

$$f_1(x + \beta x_0) = f(x) + \beta\alpha$$

is a linear functional on U_1 and it extends U. We will now show that an $\alpha \in \mathbb{R}$ can be chosen so as to assure that $\|f_1\| = 1$. By the definition of the norm $\|f_1\|$, it suffices to guarantee that

$$|f(x) + \beta\alpha| \leq \|x + \beta x_0\|,$$

for all $x \in U$ and $\beta \in \mathbb{R}$. Stated differently (replace x by $-\beta x$ and divide both sides by $|\beta|$), we seek an α such that

$$|f(x) - \alpha| \leq \|x - x_0\|,$$

for all $x \in U$. We thus wish to find an $\alpha \in \mathbb{R}$ with

$$f(x) - \|x - x_0\| \leq \alpha \leq f(x) + \|x - x_0\|,$$

for all $x \in U$. In other words, α is common to all the intervals $[A_x, B_x]$, for all $x \in U$, where $A_x = f(x) - \|x - x_0\|$ and $B_x = f(x) + \|x - x_0\|$. The existence of such an α will follow by showing that $A_x \leq B_y$ for all $x, y \in U$. And indeed,

$$B_y - A_x = f(y) - f(x) + \|y - x_0\| + \|x - x_0\|$$

and since

$$|f(y) - f(x)| = |f(y - x)| \leq \|y - x\| \leq \|y - x_0\| + \|x - x_0\|$$

it follows that

$$B_y - A_x \geq 0.$$

We thus established that f, and, importantly, that in fact any linear operator defined on a proper subspace, can be extended to the span of its current domain with one additional vector added, without altering the norm.

The stage is now clear for an application of Zorn's Lemma to further extend f_1, one dimension at a time, until we arrive at a linear operator whose domain is the entire space V. Let P be the collection of all pairs (U', f') where $U' \supseteq U$ is a linear subspace of V and $f' : U' \to \mathbb{R}$ is a linear functional extending f and satisfying $\|f'\| = 1$. Introduce a partial order on P by means of

$$(U', f') \leq (U'', f'')$$

precisely when $U' \subseteq U''$ and when f'' extends f'. The fact that P is then a poset, is immediate. We must now establish the conditions of Zorn's Lemma. Firstly, P is non-empty since (U, f) is a member of P. Next, given a chain $\{(U_i, f_i)\}_{i \in I}$ in P, consider $U' = \bigcup_{i \in I} U_i$ and define $f' : U' \to \mathbb{R}$ as follows. Given $u \in U'$ there exists an index $i \in I$ with $u \in U_i$. If $j \in I$ is another index with $u \in U_j$, then, since we are given a chain, either f_i extends f_j or f_j extends f_i. In either case setting $f'(u) = f_i(u) = f_j(u)$ produces a well-defined function. The function f' is a linear functional. Indeed, if $u_1, u_2 \in U'$, then $u_1 \in U_{i_1}$ and $u_2 \in U_{i_2}$ for some $i_1, i_2 \in I$. But, by the chain condition, we may assume without loss of generality that $U_{i_1} \subseteq U_{i_2}$. Thus

$$f'(u_1 + u_2) = f_{i_2}(u_1 + u_2) = f_{i_2}(u_1) + f_{i_2}(u_2) = f'(u_1) + f'(u_2)$$

and thus f' is additive. A similar argument shows that f is also homogenous. We leave it to the reader to similarly show that $\|f'\| = 1$ and thus it is now clear that (U', f') is in P and is an upper bound for the given chain. Note however that there is no guarantee that f' is the desired linear functional F, since U' may be properly contained in V.

With the conditions of Zorn's Lemma now verified, the existence of a maximal element $(U_M, F) \in P$ is guaranteed. If $U_M = V$, then F is a linear functional on all of V, it extends f, and its norm is 1, as required. Assume thus that U_M is properly contained in V, and let $x_0 \in V - U_M$. But, at the beginning of the proof we saw that in such a situation F can be extended to the linear subspace $U_!$, the span of $U_M \cup \{x_0\}$, without altering the norm. That would give rise to an element $(U_!, f_!) \in P$ with $(U_M F) < (U_!, f_!)$, an impossibility. The proof is thus complete. □

There are numerous consequence of the Hahn-Banach Theorem, of which we only explore a few.

Theorem 5.20 *Given a normed space V and $x_0 \in V$ with $x_0 \neq 0$, there exists a linear functional $F \in V^*$ such that $\|F\| = 1$ and $F(x_0) = \|x_0\|$.*

Proof Let

$$V_0 = \{\alpha x_0 \mid \alpha \in \mathbb{C}\}$$

be the subspace of V spanned by x_0 and let $F_0 : V_0 \to \mathbb{C}$ be given by

$$F_0(\alpha x_0) = \alpha \|x_0\|.$$

Clearly, F_0 is a linear functional, $\|F_0\| = 1$, and $F_0(x_0) = \|x_0\|$. By the Hahn-Banach Theorem, an extension $F \in V^*$ exists, establishing the claim. □

Corollary 5.5 *For all $x \in V$*

$$\|x\| = \sup_{F \in V^*, \|F\|=1} |F(x)|.$$

Corollary 5.6 *For all vectors x in a normed space V, if $F(x) = 0$ for all $F \in V^*$, then $x = 0$.*

Theorem 5.21 *Let V be normed space and U a linear subspace of it. Given $x_0 \in U$, it holds that x_0 is in the closure of U if, and only if, there exists no bounded linear functional f on V with the property that $f(x) = 0$ for all $x \in U$ but $f(x_0) \neq 0$.*

Proof Suppose that x_0 is in the closure of U. There exists then a sequence $\{u_n\}_{n \geq 1}$ of element in U converging to x_0. Remembering that a bounded linear functional $f : V \to K$ is in particular continuous, it follows that $f(u_n) \to f(x_0)$. But then, if $f(x) = 0$ for all $x \in U$, it follows that $f(x_0) = 0$. In the other direction, suppose x_0 is not in the closure of U. Then there exists a positive δ such that $\|x - x_0\| > \delta$, for all $x \in U$. Consider the span U' of the set $U \cup \{x_0\}$. Define now $f_1(x + \beta x_0) = \beta$ (remember that the elements of the span are of the form $x + \beta x_0$, with $\beta \in K$ and $x \in U$). This is clearly a linear functional on U_1 and since

$$\delta|\beta| \leq |\beta| \|x_0 + \frac{1}{\beta} x\| = \|\beta x_0 + x\|$$

we have that $\|f_1\| \leq 1/\delta$. Since, by definition, $f_1(x) = 0$ for all $x \in U$ and $f_1(x_0) = 1$, applying the Hahn-Banach Theorem to extend f' to all of V establishes the claim. □

Exercises

Exercise 5.31 Let V be a normed space. Prove that V^* is bounded if, and only if, $\dim(\mathrm{V}) = 0$.

Exercise 5.32 Apply the Riesz Representation Theorem to $\mathbb{R}$ and $\mathbb{C}$, each with its standard inner product, and give an explicit construction of the function ψ, witnessing its stated properties.

Exercise 5.33 Let V be a normed space and $f \in V^*$. Prove that $p_f : V \to \mathbb{R}$ given by $p_f(x) = |f(x)|$ is a semi-norm on V. Under which conditions on V can it be a norm?

Exercise 5.34 Complete the proof of the fact that, as real spaces, $\ell_1^* \cong \ell_\infty$.

Exercise 5.35 Does $V^* \cong V$ hold for general normed spaces?

Exercise 5.36 Let $\{x_1, \ldots, x_n\}$ be n linearly independent vectors in a normed space V, and $z_1, \ldots, z_n \in \mathbb{C}$. Prove the existence of a linear functional $F \in V^*$ with

$$F(x_k) = z_k$$

for all $1 \le k \le n$.

Exercise 5.37 Consider the space $C([a, b], \mathbb{R})$ with the L_∞ norm. Given an integrable function $y : [a, b] \to \mathbb{R}$, prove that the function $F_y : C([a, b], \mathbb{R}) \to \mathbb{R}$ given by

$$F_y(x) = \int_a^b dt\ y(t)x(t)$$

is a linear functional and show that

$$\|F\| = \int_a^b dt\ |y(t)|.$$

Exercise 5.38 Show that the extension guaranteed by the Hahn-Banach Theorem is, generally speaking, far from unique by presenting a linear operator with infinitely many extensions. Proceed in two ways, one by analyzing the proof of the Hahn-Banach Theorem, the other by giving explicit extensions of a well-chosen linear operator (hint: consider finite dimensional spaces).

Exercise 5.39 In the previous exercise you constructed explicit extensions of a linear functional. Convince yourself that for infinite dimensional linear spaces it may be very hard, if at all possible, to obtain an explicit extension of an arbitrary functional.

Exercise 5.40 Use the Hahn-Banach Theorem to prove that it is possible to assign to every bounded sequence $\{x_m\}_{m\ge1}$ of real numbers a real number $\lim_{m\to\infty} x_m$ such that

1. If $\{x_m\}$ is a convergent sequence, then $\lim_{m\to\infty} x_m$ is the limit of the sequence in the usual sense.
2. $\lim_{m\to\infty}(x_m + y_m) = \lim_{m\to\infty} x_m + \lim_{m\to\infty} y_m$ for all bounded sequences $\{x_m\}_{m\ge1}$ and $\{y_m\}_{m\ge1}$.
3. $\lim_{m\to\infty} \alpha x_m = \alpha \lim_{m\to\infty} x_m$ for all bounded sequences $\{x_m\}_{m\ge1}$ and all scalars $\alpha \in \mathbb{R}$.

Can you determine $\lim_{m\to\infty}(-1)^m$?

5.5 Unbounded Operators and Locally Convex Spaces

The theory that was presented so far is primarily concerned with the theory of bounded linear operators between normed spaces. In this section we present two directions for generalizations of the theory. The first one stems from the observation made above to the effect that while bounded operators encompass a lot of examples of interest, some important operators, the differentiation operator for instance, are not bounded. One is thus compelled to consider unbounded operators. The second generalization we consider originates from the fact that at times the space one is considering does not support a norm. It turns out that if instead of a norm one has a suitable family of semi-norms, then one can still recover much of the general theory. Spaces with such a family of semi-norms are called locally convex spaces, and we will only investigate the definition and some illustrative examples, without delving deeper into the general theory.

5.5.1 *Closed Operators*

The development of the theory of unbounded linear operators was motivated by the unbounded nature of the differentiation operator, as well as the development of a mathematical framework for Quantum Mechanics (e.g., to account for unbounded observables). Arbitrary unbounded operators may be too wild to tame, and so some restrictions must be placed in order to identify a suitable class of operators for which a reasonable theory emerges. The closed operators, presented below, form a class of operators broader than the bounded ones admitting a strong general theory that allows one to analyze situations that lie out of the rich of the theory of bounded operators. The results we present below are only the tip of the unbounded iceberg.

In the context of unbounded linear operators it is common to re-interpret the definition of an operator between linear spaces to be one that is not necessarily defined on all of the specified domain.

Definition 5.12 Let $\mathscr{B}_1$ and $\mathscr{B}_2$ be Banach spaces. An *(unbounded) linear opeartor* $A : V \to W$ consists of a linear subspace $\mathscr{D}(A) \subseteq V$ and a function $A : \mathscr{D}(A) \to W$ which is additive and homogenous. Further, A is a *closed linear operator* when the following condition holds. For every sequence $\{x_m\}_{m \geq 1}$ in V if $x_m \to x_0$ for some $x_0 \in V$ and $Ax_m \to y_0$ for some $y_0 \in W$, then $x_0 \in \mathscr{D}(A)$ and $Ax_0 = y_0$.

Example 5.16 In Example 5.4 we already saw that the differentiation operator $d/dt : C^1([0, 1], \mathbb{R}) \to C([0, 1], \mathbb{R})$, with the L_∞ norm, is an unbounded linear operator. In the context of unbounded operators we may thus consider

$$d/dt : C([0, 1], \mathbb{R}) \to C([0, 1], \mathbb{R}),$$

with the L_∞ norm, where we may specify various different subspaces as the domain of definition. Different choices for the actual domain may lead to completely different properties of the operator. For instance, if we take the domain to be $C^1([a, b], \mathbb{R})$, then we obtain an unbounded closed operator. Indeed, this claim follows from the elementary results that if $\{x_m\}_{m\geq 1}$ is a sequence of functions that converge (pointwise) to x_0, and their derivatives x_m' converge uniformly to y_0, then x_0 is differentiable and $x_0' = y_0$. In contrast, taking the domain of d/dt to be $C^\infty([0, 1], \mathbb{R})$ yields a non-closed operator.

The defining condition of a closed operator $A : \mathscr{B}_1 \to \mathscr{B}_2$ can be restated in terms of its *graph* $\Gamma(A) = \{(x, Ax) \in \mathscr{B}_1 \times \mathscr{B}_2 \mid x \in \mathscr{D}(A)\}$, as follows.

Proposition 5.5 *An unbounded linear operator $A : \mathscr{B}_1 \to \mathscr{B}_2$ between Banach spaces is closed if, and only if, the graph $\Gamma(A)$ is closed in $\mathscr{B}_1 \times \mathscr{B}_2$.*

The proof is left for the reader. We now prove that a closed linear operator is automatically bounded, provided its domain is closed.

Theorem 5.22 (Banach's Closed Graph Theorem) *Let $A : \mathscr{B}_1 \to \mathscr{B}_2$ be a closed linear operator between Banach spaces. If the domain $\mathscr{D}(A)$ is closed in $\mathscr{B}_1$, then A is bounded.*

Proof First, it is straightforward to verify that the space $\mathscr{B}_1 \times \mathscr{B}_2$ with norm given by

$$\|(x, y)\| = \|x\| + \|y\|$$

is a Banach space. Since A is closed, its graph $\Gamma(A) \subseteq \mathscr{B}_1 \times \mathscr{B}_2$ is thus a closed subset of a complete space, and thus is itself a Banach space. Similarly, by assumption, $\mathscr{D}(A)$ is closed in $\mathscr{B}_1$ and thus is a Banach space too.

We now consider the operator $P : \Gamma(A) \to \mathscr{D}(A)$ given by

$$P(x, Ax) = x,$$

clearly a linear operator. Since

$$\|P(x, Ax)\| = \|x\| \leq \|x\| + \|Ax\| = \|(x, Ax)\|$$

we see that P is a bounded linear operator. Moreover, P is invertible (its inverse P^{-1} obviously given by $x \mapsto (x, Ax)$) and thus, by Corollary 5.2, P^{-1} is bounded, say by M. Therefore,

$$\|P^{-1}x\| = \|x\| + \|Ax\| \leq M\|x\|$$

for all $x \in \mathscr{D}(A)$, and thus

$$\|Ax\| \leq (M - 1)\|x\|$$

showing A is bounded. □

We note that this is yet another deep result in the theory of Banach spaces, seeing that this rather short proof uses a corollary of the Open Mapping Theorem.

5.5.2 Locally Convex Spaces

As motivation for the concept we now present, consider the set $\mathbb{R}^\infty$ of all sequences of real numbers (with obvious modifications one may also consider $\mathbb{C}^\infty$). Of course the linear structure of $\mathbb{R}$ carries over to $\mathbb{R}^\infty$ by point-wise operations and thus $\mathbb{R}^\infty$ is a linear space. However, there is no natural way to turn $\mathbb{R}^\infty$ into a normed space. In particular, each ℓ_p norm is only defined on a proper subset of $\mathbb{R}^\infty$ and can not be extended to a norm on $\mathbb{R}^\infty$. There is however a natural choice of a family of semi-norms, defined as follows. For each $k \in \mathbb{N}$ let $p_k : V \to \mathbb{R}$ be given by $p_k(x) = |x_k|$, the absolute value of the k-th component of x. It is easily seen that p_k is indeed a semi-norm.

We thus see that in the absence of a norm a space may still admit a family of semi-norms and it is the case that, under suitable conditions, significant portions of the theory developed above (in particular the Hahn-Banach Theorem) have suitable analogues in locally convex spaces. We will now present the definition and explore some of the most basic aspects of these spaces.

Definition 5.13 A *locally convex linear space* is a linear space V together with a family $\{p_i\}_{i\in I}$ of semi-norms on V. The family of semi-norms is said to *separate points* if from $p_i(x) = 0$ for all $i \in I$ it follows that $x = 0$.

Example 5.17 Of course any normed space and any semi-normed space is a locally convex space. It is not hard to show that any locally convex space given by a finite family of semi-norms is in fact semi-normed. Thus the concept of locally convex spaces becomes substantial only when once considered infinite families of semi-norms, as in the case of $\mathbb{R}^\infty$ above.

We now consider the topology induced by the family of semi-norms in a locally convex space.

Definition 5.14 Let V be a locally convex space given by the family $\{p_i\}_{i\in I}$ of semi-norms. For each $i \in I$ and $y \in V$ let $f_{i,y} : V \to \mathbb{R}$ be the function

$$f_{i,y}(x) = p_i(x - y).$$

The *induced topology* on V is the smallest topology such that each of $f_{i,y} : V \to \mathbb{R}$ is continuous.

The existence of such a smallest topology is guaranteed by Proposition 3.11. In fact, from the proof of that proposition, a local base at y for the induced topology is given by the collection $U_{B,\varepsilon}$, where $B \subseteq \{p_i\}_{i\in I}$ is a finite subset of the given family of semi-norms and $\varepsilon > 0$, and

$$U_{B,\varepsilon} = \{x \in V \mid p_i(x - y) < \varepsilon \quad \forall i \in B\}.$$

The following result now follows easily, and thus the proof is left for the reader.

Proposition 5.6 *Let V be a locally convex space given by the family $\{p_i\}_{i\in I}$ of semi-norms.*

1. *A sequence $\{x_m\}_{m\geq 1}$ in V converges to $x_0 \in V$ in the induced topology if, and only if, $p_i(x_m - x_0) \to 0$ for all $i \in I$.*
2. *The induced topology is Hausdorff if, and only if, the family of semi-norms separates points.*
3. *The linear space operations are continuous.*

With these observations made, the resemblance between the theory of normed spaces we developed above and the theory of locally convex spaces with separated points is only starting to become visible. Without a doubt, it is more convenient to have a single norm and appeal to the rich theory of normed spaces whenever possible. However, often enough a norm is simply not available. The theory of locally convex spaces is rich enough so as to justify the somewhat cumbersome management of a family of semi-norms instead of a single norm. As mentioned above, significant portions of the theory of (semi-)normed spaces transfers quite smoothly to the theory of locally convex spaces, however we do not explore this any further.

Exercises

Exercise 5.41 Prove that if $\mathscr{B}_1$ and $\mathscr{B}_2$ are Banach spaces, then so is $\mathscr{B}_1 \times \mathscr{B}_2$ with norm given by $\|(x, y)\| = \|x\| + \|y\|$.

Exercise 5.42 Prove that if $A : \mathscr{B}_1 \to \mathscr{B}_2$ is a closed linear operator, then its kernel $\text{Ker}(A)$ is a closed subspace of $\mathscr{B}_1$.

Exercise 5.43 Prove that any bounded linear operator is a closed operator.

Exercise 5.44 Prove Proposition 5.5.

Exercise 5.45 Let V be a normed space. We may now consider an induced topology on it in two ways. Namely, we may view it as a locally convex space where the family of semi-norms consists of just the given norm on V, and obtain the induced topology, or we may consider the induced metric $d(x, y) = \|x - y\|$ and consider its induced topology. Show that the two topologies coincide.

Exercise 5.46 Prove that every locally convex linear space V defined using a finite family $\{p_i\}_{i\in I}$ of semi-norms is semi-normed. In more detail, construct a single semi-norm on V which induces the same topology as the finite family of semi-norms does.

Exercise 5.47 Let V be a locally convex space given by the countable family $\{p_i\}_{i\geq 1}$ of semi-norms. Consider the function $d : V \times V \to \mathbb{R}$ given by

$$d(x, y) = \sum_{i=1}^{\infty} \frac{1}{2^n} \frac{p_i(x - y)}{p_i(x - y) + 1}.$$

Prove that (V, d) is a semimetric space, and that it is a metric space if, and only if, the family of norms separates points.

Exercise 5.48 Continuing the previous exercise, show that the topology induced by the semimetric and by the family of semi-norms coincide. (You will need to define the topology induced by a semimetric.)

Exercise 5.49 Consider $\mathbb{R}$ with the standard structure of an inner product space and let $V = C(\mathbb{R}, \mathbb{R})$ be the linear space of all continuous functions $x : \mathbb{R} \to \mathbb{R}$. For each $t \in \mathbb{R}$ let

$$p_t(x) = |x(t)|.$$

For each $n \in \mathbb{N}$ let

$$q_n(x) = \max_{-n \leq t \leq n} |x(t)|.$$

For each compact subset $C \subseteq \mathbb{R}$ let

$$r_C(x) = \max_{t \in C} |x(t)|.$$

For each of the families $\{p_t\}_{t \in \mathbb{R}}$, $\{q_n\}_{n \in \mathbb{N}}$, and $\{r_C\}_{C \subseteq \mathbb{R}}$, where C ranges over the compact subsets of $\mathbb{R}$, decide whether it endows V with the structure of a locally convex space, whether the family of semi-norms separates points, and compare the induced topologies.

Exercise 5.50 Prove Proposition 5.6.

Further Reading

The hitchhiker's guide to infinite dimensional analysis ([1]) offers a comprehensive source on all the topics covered in this chapter and far beyond. For a more concise introduction to Banach spaces, complete with a detailed preliminaries chapter, see [2]. For a classic introduction to functional analysis see [5], and for a text somewhat different in style, including an entire chapter devoted to unbounded operators, see [4]. To read more about locally convex spaces the reader may consult [3].

References

1. C.D. Aliprantis, K.C. Border, *Infinite Dimensional Analysis (A Hitchhiker's Guide)*, Third edn. (Springer, Berlin, 2006), p. 703
2. N.L. Carothers, *A Short Course on Banach Space Theory, London Mathematical Society Student Texts* (Cambridge University Press, Cambridge, 2005), vol. 64, p. 184
3. M.S. Osborne, *Locally Convex Spaces, Graduate Texts in Mathematics* (Springer, Cham, 2014), vol. 269, p. 213
4. G.K. Pedersen, *Analysis Now, Graduate Texts in Mathematics* (Springer, New York, 1989), vol. 118, p. 277
5. K. Yosida, *Functional Analysis, Classics in Mathematics* (Springer, Berlin, 1995), p. 501

Chapter 6
Topological Groups

Abstract This short chapter is a brief introduction to the theory of topological groups, set in the context of Banach space theory. Not assuming any knowledge of groups, the chapter is entirely self-contained, presenting the definitions of groups, homomorphisms, (normal) subgroups, and quotient groups before moving on to the definition of topological groups. Some fundamental consequences of the interaction between algebra and topology are discussed. Introducing uniform spaces, the main message of the chapter is that the topology of a topological group is nearly metrizable, in the sense that it is uniformizable.

Keywords Group · Group homomorphism · Topological group · Uniformity · Quotient group · Subgroup · Normal subgroup · Abelian group

Banach spaces admit a particularly rich theory since the algebraic structure (i.e., the norm) induces a particularly nice metric (i.e., a complete one) resulting in a powerful interaction between algebra and geometry. Quite often, when an algebraic structure and a topological structure are allowed to interact, the fusion of the two theories results in a very intricate and interesting new theory. Such a fusion is the focus of this chapter where we present topological groups.

There are at least two motivating reasons for considering topological groups. The first one is that just as groups model symmetry, so do topological groups model continuous symmetry. The second reason for studying topological groups is that quite often a portion of, say, a Banach space may fail to be a Banach space on its own but it may still retain some of the algebra and the topology of the ambient space to form a topological group. This is often the case since a group is a much weaker algebraic structure compared to a linear space, and a topology is much weaker than a metric (whether or not induced by a norm). Topological groups thus are much more common-place since one only requires a group structure and a topology, rather than a full linear space and a metric structure, yet the theory of topological groups still presents a powerful fusion between algebra and geometry.

Section 6.1 introduces groups and group homomorphisms, without assuming any prior knowledge of groups. The main objects of study, namely topological groups, are then presented in Sect. 6.2. Section 6.3 presents topological subgroups and a first encounter with the interesting interaction between algebra and topology, namely that the closure of a subgroup is a subgroup. The quotient construction for topological

© Springer International Publishing Switzerland 2015

C. Alabiso and I. Weiss, *A Primer on Hilbert Space Theory*,
UNITEXT for Physics, DOI 10.1007/978-3-319-03713-4_6

groups is the topic of Sect. 6.4, including a rather detailed account of normal subgroups. Finally, Sect. 6.5 discusses another, deeper, consequence of the interaction between algebra and topology, namely the uniformizability of the topology of a topological group, thus enabling one to import significant portions of the uniform machinery of metric spaces into the realm of topological groups.

6.1 Groups and Homomorphisms

The axioms defining a group are an abstraction of structure present in many familiar algebraic systems, namely an associative and unital binary operation with respect to which every element is invertible. The historical development of group theory is rooted in the study of roots of polynomials, famously with the work of Evariste Galois and Niels Henrik Abel on finite groups. There is a distinct difference between the theory of finite groups and the theory of infinite groups. In finite group theory various counting arguments play an important role in determining the structure of groups. The recent classification of all finite simple groups, the result of an almost unfathomable collaboration spanning thousands of articles, is a milestone of finite group theory. Infinite groups though can be so wild that a full classification of all groups seems beyond reasonable expectations. Some limitations on size must be placed, and one way to impose such size constraints is through the introduction of a topology, and demanding that the group, as a topological space, be compact. The classification of compact Hausdorff topological groups is a much more manageable project.

With that in mind, we turn now to present groups, but with an eye towards their common applications in Physics (i.e., as tools for the study of symmetry) and as a step towards defining topological groups (and thus finite groups are only glanced at).

Definition 6.1 A *group* is a triple $(G, \cdot, e)$ where G is a set, $\cdot$ is a binary operation $G \times G \to G$, usually denoted by $(g, h) \mapsto g \cdot h$, or even just $(g, h) \mapsto gh$, and $e \in G$ is a chosen element, such that the following axioms hold.

1. Associativity, i.e., $(g_1 g_2) g_3 = g_1 (g_2 g_3)$ for all $g_1, g_2, g_3 \in G$.
2. The element e is an *identity* element, i.e., $eg = g = ge$ for all $g \in G$.
3. Existence of *invereses*, i.e., for all $g \in G$ there exists an element, denoted by $g^{-1} \in G$ and called the *inverse* of g, satisfying $gg^{-1} = e = g^{-1}g$.

If $gh = hg$ holds for all $g, h \in G$, then the group is said to be *commutative* or *abelian*. In an abelian group it is customary to write $g + h$ instead of gh, and to denote e, the neutral element, by 0. The cardinality of G is called the *order* of the group.

When there is no danger of confusion it is common to refer to a group G, leaving the operation $\cdot$ and the identity e implicit.

Proposition 6.1 *In a group G*

1. *the identity is unique, i.e., if $e' \in G$ has the property that $e'g = g = g'e$ for all $g \in G$, then $e' = e$;*
2. *inverses are unique, i.e., for every element $g \in G$, if $h_1, h_2 \in G$ satisfy*

$$gh_2 = gh_1 = e = h_1 g = h_2 g,$$

then $h_1 = h_2$.

Proof The arguments are very similar to those given in the proof of Proposition 2.1, and thus we leave it to the reader to adapt that proof to the current situation. □

Remark 6.1 An inspection of the definition of linear space (Definition 2.1) reveals that we have already encountered abelian groups. Indeed, in that definition there are three sets of axioms, the first of which states, precisely, that $(V, +, 0)$ is an abelian group for any linear space V. Thus, had we chosen to present groups prior to linear spaces, we could have defined a linear space as an abelian group together with a scalar product, satisfying the other two sets of axioms in the definition of linear space.

Many examples of groups arise as the set of symmetries of an object. For instance, fixing an equilateral triangle in the plane, its symmetries correspond to the various ways in which rigid motions of the plane map the triangle onto itself. The set of all rigid motions that map the triangle to itself (of which there are precisely six: the identity, three reflections, and two rotations) is a group under the operation of composition. The verification is straightforward: the composition of rigid motions that fix the triangle is again a rigid motion that fixes the triangle, the identity function is a rigid motion that fixes the triangle (and thus serves as the identity element of the group), and the inverse function of a rigid motion that fixes the triangle is again a rigid motion that fixes the triangle. The resulting group is called the *dihedral group* of order 6, one member in the following family of groups.

Theorem 6.1 *Let P be a regular n-gon in the plane (i.e., a regular polygon with n vertices), $n \geq 2$. The set D_n of all rigid motions (i.e., functions of the plane to itself that preserve lengths and angles) together with the operation of composition of functions is a group.*

Proof The reader is invited to turn the argument for D_3 given above into a proof of the general case. □

Remark 6.2 The groups $\{D_n\}_{n \geq 2}$ are known as the *dihedral groups*. It can be shown that D_n has order $2n$. For that reason, the dihedral group D_n is also commonly (and confusingly) denoted by D_{2n}.

Of course, one may vary the shape being fixed, as well as the degree to which the mappings preserve the ambient geometry, or change the ambient geometry itself. As an extreme example, given a set S, the set of all bijections $\sigma : S \to S$, together with composition of functions, is easily seen to be a group. This is the group of symmetries of the set S, clearly essentially determined only by the cardinality of S. This group is denoted by $\mathrm{Sym}(S)$, or, when $S = \{1, 2, 3, \ldots, n\}$, by S_n.

Following the same general idea, the following is an important example in the theory of topological groups.

Theorem 6.2 *For $n \geq 1$ let $\mathrm{GL}_n(\mathbb{R})$ be the set of all invertible linear operators $A : \mathbb{R}^n \to \mathbb{R}^n$. With the operation of composition of functions, $\mathrm{GL}_n(\mathbb{R})$ is a group.*

Proof The straightforward verification of the axioms, using established results from Chap. 2, is left for the reader. □

The group $\mathrm{GL}_n(\mathbb{R})$ is known as the *general linear group*. It is the group of linear symmetries of the linear space $\mathbb{R}^n$. By specifying a basis for $\mathbb{R}^n$ one may identify $\mathrm{GL}_n(\mathbb{R})$ with the set of all invertible $n \times n$ matrices with real entries. The group $\mathrm{GL}_n(\mathbb{R})$ is essentially the same as the group of invertible $n \times n$ matrices over $\mathbb{R}$ with the operation of matrix multiplication. Other important groups are defined in terms of certain linear symmetries, or, equivalently, as spaces of matrices satisfying various conditions. See the solved problems section for this chapter for detailed examples.

As a matter of fact, the claim above is a special case of a much more general source of groups, as we now show.

Theorem 6.3 *Let $\mathscr{B}$ be a Banach space. The subset G of $\mathbf{B}(\mathscr{B})$ consisting of all invertible bounded linear operators on $\mathscr{B}$, with the operation of composition of functions, is a group.*

Proof The implicit claim that the composition of invertible bounded linear operators is again an invertible bounded linear operator follows from the fact that

$$\|A_1 A_2\| \leq \|A_1\| \|A_2\|$$

for all operators $A_1, A_2 \in \mathbf{B}(\mathscr{B})$, as is easily established. Next, clearly, the identity $\mathrm{id}_{\mathscr{B}}$ is a member of $\mathbf{B}(\mathscr{B})$, and it serves as the identity element of the group. The fact that composition of functions is associative is trivially verified. The fact that the inverse operator of a bounded operator is bounded is Corollay 5.2 of the Open Mapping Theorem. □

We now turn to the structure preserving mappings between groups.

Definition 6.2 A function $\psi : G \to H$ between groups is a *group homomorphism* (or simply a *homomorphism)* if

$$\psi(g_1 g_2) = \psi(g_1)\psi(g_2)$$

for all $g_1, g_2 \in G$. If ψ is also bijective, then it is called an *isomorphism*, and then the groups G and H are said to be *isomorphic*, denoted by $G \cong H$.

Example 6.1 As said above, in the presence of a basis for $\mathbb{R}^n$, there is a bijection between $\mathrm{GL}_n(\mathbb{R})$ and $\mathrm{GL}(n;\mathbb{R})$, given by mapping an invertible linear operator to its representing matrix. In fact, this correspondence is a group isomorphism between the two groups. The determinant function $\det : \mathrm{GL}_n(\mathbb{R}) \to \mathbb{R}$, which assigns to a linear operator $A : \mathbb{R}^n \to \mathbb{R}^n$ its determinant (i.e., the determinant of its representing matrix in any basis) is a group homomorphism $\det : \mathrm{GL}_n(\mathbb{R}) \to \mathbb{R}^*$. Here $\mathbb{R}^*$ is the set of non-zero real numbers and the group operation is ordinary multiplication. When $n = 1$ the determinant $\det : \mathrm{GL}_1(\mathbb{R}) \to \mathbb{R}^*$ is a group isomorphism.

A group homomorphism is only required to preserve the group operations in the domain and codomain. It follows immediately though that the rest of the group structure is also preserved. For the proof it is helpful to note that the existence of inverses in a group immediately implies the left and right cancelation laws: in a group G, if $gh = gh'$, then $h = h'$, and, similarly, if $hg = h'g$, then $h = h'$.

Proposition 6.2 *Let $\psi : G \to H$ be a group homomorphism. Then $\psi(e) = e$ and $\psi(g^{-1}) = \psi(g)^{-1}$, for all $g \in G$.*

Proof We only establish that $\psi(e) = e$ (notice that the e on the left is the identity in G while on the right it is the identity in H, and thus they may be different). Notice that $\psi(e)\psi(e) = \psi(ee) = \psi(e) = e\psi(e)$. The cancelation law now implies that $\psi(e) = e$. The verification that $\psi(g^{-1}) = \psi(g)^{-1}$ is now trivial. □

The verification of the following properties of homomorphisms and isomorphisms is very similar to the analogous results established for linear operators, for instance in Proposition 2.8. We thus omit the details of the proof.

Proposition 6.3 *Let $G_1 \xrightarrow{\varphi} G_2 \xrightarrow{\psi} G_3$ be group homomorphisms. Then*

1. *The composition $\psi \circ \varphi$ is a homomorphism.*
2. *If both φ and ψ are isomorphisms, then so is $\psi \circ \varphi$.*
3. *If ψ is an isomorphism, then so is ψ^{-1}.*
4. *The identity function $\mathrm{id}_G : G \to G$ is a group isomorphism.*

6.2 Topological Groups and Homomorphism

A topological group is a group together with a topology on it which is required to be compatible with the group structure, in the sense that the group operations are to be continuous. Some of the far-reaching consequences resulting from this fusion between algebra and topology will be explored in the subsequent sections. In particular, it will be shown that the topology must be sufficiently metric-like so as to render such concepts as uniform continuity and Cauchy sequences meaningful. For now we focus on the definition and examples.

Definition 6.3 A *topological group* is a group G together with a topology on the set G such that the group operation $\cdot : G \times G \to G$ is continuous (with respect to the given topology on the codomain and the product topology on the domain), and such that the function $G \to G$ given by $g \mapsto g^{-1}$ is continuous.

Example 6.2 The following examples of topological groups are all subsets of $\mathbb{R}$ or $\mathbb{C}$, and we consider each of them with the subspace topology induced, in each case, by the metric $d(x, y) = |x - y|$.

1. The circle $S^1 = \{z \in \mathbb{C} \mid |z| = 1\}$ with the operation of multiplication.
2. The set $\mathbb{R}$ of all real numbers with the operation of addition.
3. The set $\mathbb{R}^*$ of all non-zero real numbers with the operation of multiplication.
4. The set $\mathbb{C}$ of all complex numbers with the operation of addition.
5. The set $\mathbb{C}^*$ of all non-zero complex numbers with the operation of multiplication.

A large family of examples of topological groups are obtained by the following result.

Lemma 6.1 *Let V be a normed space. Then $(V, +, 0)$, i.e., the additive structure of the space, is a topological group.*

Proof We already noted that the additive part of a linear space is an abelian group. The continuity requirements are satisfied by Proposition 5.1. □

Another significant source of topological groups arises from groups of bounded linear operators on a Banach space.

Theorem 6.4 *Let $\mathscr{B}$ be a Banach space. The subset G of $\mathbf{B}(\mathscr{B})$ consisting of the invertible bounded linear operators on $\mathscr{B}$ is a topological group when endowed with the operator norm topology.*

Proof Having already observed that G is a group under composition of functions, it remains to verify the continuity of the composition and the inverse mapping with respect to the operator norm. The continuity of the composition follows at once from the inequality

$$\|A_1 A_2\| \leq \|A_1\| \|A_2\|$$

while the continuity of the inverse mapping is the statement of Exercise 5.30. □

Corollary 6.1 *The general linear group* $\mathrm{GL}_n(\mathbb{R})$, *when endowed with the operator norm topology, is a topological group (which is not abelian if $n > 1$).*

Proof $\mathrm{GL}_n(\mathbb{R})$ is the group of invertible operators on the Banach space $\mathbb{R}^n$. □

Since a topological group is a fusion between a group and a topological space, so are homomorphisms of topological groups a fusion between group homomorphisms and continuous mappings.

Definition 6.4 A function $\psi : G \to H$ between topological groups is said to be a *homomorphism* if ψ is both continuous and a group homomorphism. We say that ψ is an *isomorphism* if ψ is both a group isomorphism and a homeomorphism. If an isomorphism between topological groups G and H exists, then the groups are said to be *isomorphic*, denoted by $G \cong H$.

Example 6.3 In a topological group G, the mapping $g \mapsto g^{-1}$ is a homeomorphism but not usually a group homomorphism. Indeed, it is a group homomorphism if G is abelian, since then

$$(gh)^{-1} = g^{-1}h^{-1}$$

while in general one only has the equality

$$(gh)^{-1} = h^{-1}g^{-1}.$$

For any fiexed element $a \in G$, the *left translation* mapping $g \mapsto ag$ and the *right translation* mapping $g \mapsto ga$ are both homeomorphisms, but are generally not group homomorphisms. The determinant function $\det : \mathrm{GL}_n(\mathbb{R}) \to \mathbb{R}^*$ is continuous (since the determinant is polynomial in the entries of the matrix) and, as we have seen, is a group homomorphism, and thus is a homomorphism of topological groups. $\det : \mathrm{GL}_1(\mathbb{R}) \to \mathbb{R}^*$ is an isomorphism.

The reader is invited to state and prove the analogous result of Proposition 6.3 for topological homomorphisms and isomorphisms.

6.3 Topological Subgroups

A topological subgroup of a topological group is a subset which, on its own right, is a topological group with the induced algebraic and topological structures from the ambient topological group. In other words, a topological subgroup is a fusion of the concepts subgroup and topological subspace. Fortunately, it turns out that the subspace topology on a subgroup is automatically compatible with the group structure, and thus it is only the algebraic structure that dictates what the topological subgroups are. Consequently, there is no difference between the topological subgroups of a topological group and its subgroups when the topology is forgotten.

In light of the above, we focus on the notion of subgroup, and, anticipating the contents of the next section, already discuss normal subgroups. We then observe the first non-trivial consequence of the interaction of the algebra and the topology, namely that the closure of a subgroup is a subgroup.

Definition 6.5 A subset H of a group G is a *subgroup* of G if the group operation in G restricts to a group operation on H. Equivalently, H is a subgroup of G if $e \in H$ and for all $h_1, h_2 \in H$

$$h_1 h_2^{-1} \in H.$$

The motivation for the following definition is presented in the next section.

Definition 6.6 A subgroup H of a group G is said to be *normal* if $ghg^{-1} \in H$ for all $h \in H$ and $g \in G$. The element ghg^{-1} is called the *conjugate* of h by g. Thus, a subgroup is normal when it is closed under conjugation.

Remark 6.3 Note that the condition for normality of H in G is equivalent to: for all $g \in G$ and $h \in H$, there exists $h' \in G$ such that

$$hg = gh'.$$

Thus, the condition of normality can be seen as a weak form of commutativity; elements in G commute with elements in H, up to a replacement by another element in H. In particular, if G is abelian, then every subgroup H of G is normal.

Example 6.4 Given a group homomorphism $\psi : G \to H$, its *kernel* is the set $\text{Ker}(\psi) = \{g \in G \mid \psi(g) = e\}$. The kernel of a homomorphism is easily seen to always be a normal subgroup of G. One may also easily verify (compare with Theorem 2.8) that $\text{Ker}(\psi) = \{e\}$, i.e., the kernel is the *trivial subgroup* of G, if, and only if, ψ is injective.

Regarding subgroups of a topological group G, the presence of a topology entails no complications, at least in the following sense.

Lemma 6.2 *Let G be a topological group and H a subgroup of G. When endowed with the subspace topology, H is a topological group.*

The proof is immediate, relying on the simple fact that the restriction of a continuous mapping to a subspace remains continuous.

We may now demonstrate one consequence of the fruitful interaction between the group structure and the topology.

Theorem 6.5 *If H is a (normal) subgroup of a topological group G, then $\overline{H}$, the closure of H, is also a (normal) subgroup of G.*

Proof To show that $\overline{H}$ is a subgroup, we need to show that $gh^{-1} \in \overline{H}$ for all $g, h \in \overline{H}$. In other words, for the function $f : G \times G \to G$ given by $f(g, h) = gh^{-1}$, we need to establish that $f(\overline{H} \times \overline{H}) \subseteq \overline{H}$. Since f is clearly continuous, and $\overline{H}$ is closed, it follows that $f^{-1}(\overline{H})$ is closed. Next, $H \times H \subseteq f^{-1}(H) \subseteq f^{-1}(\overline{H})$, where the first inclusion follows from the fact that H is a subgroup and the second inclusion since $H \subseteq \overline{H}$. Taking closures leads to $\overline{H \times H} \subseteq \overline{f^{-1}(\overline{H})} = f^{-1}(\overline{H})$. The result follows by noting that $\overline{H} \times \overline{H} = \overline{H \times H}$ (this latter fact is a general property of closures that has nothing to do with the group structure on H). We leave it to the reader to verify that if H is a normal subgroup, then so is $\overline{H}$. □

6.4 Quotient Groups

Whereas the notion of subgroup of a topological group was quite straightforward, with a seamless interaction between the algebraic and the topological demands, the situation for the quotient of a topological group by a subgroup is more intricate. The main difficulty lies with the algebraic structure, namely that for an arbitrary subgroup H, it is not always possible to obtain a group structure on the relevant quotient set. Topologically however, there are no difficulties since the quotient topology is always available. The details are given below.

Given a group G and a subgroup H, defining $g_1 \sim g_2$ precisely when $g_1^{-1}g_2 \in H$ is easily seen to be an equivalence relation on G. The equivalence class of $g \in G$ is easily seen to be the set

$$gH = \{gh \mid h \in H\}$$

and is called the *left coset of* g. In particular, two left cosets g_1H and g_2H are equal if, and only if, $g_1^{-1}g_2 \in H$. Similarly, defining $g_1 \sim g_2$ precisely when $g_1g_2^{-1} \in H$ gives rise to an equivalence relation whose equivalence classes are all of the form

$$Hg = \{hg \mid h \in H\}$$

and are called *right cosets* (of H in G). Of course, if G is abelian, then the left and right cosets coincide, but in general they may be different. However, any general result concerning left cosets in a group has a corresponding dual result about right cosets. In fact the translation between results on left and right cosets is achieved by means of the correspondence $gH \leftrightarrow Hg^{-1}$. For this reason, we may safely only concentrate on either left or right cosets, and we choose the left ones.

Thus, any choice of a subgroup H of G gives rise to an equivalence relation $\sim$ with corresponding quotient set

$$G/H = \{gH \mid g \in G\},$$

the set of all left cosets of H in G, with the corresponding canonical projection $\pi : G \to G/H$ given by $\pi(g) = gH$. We now address the question whether the group structure on G induces a group structure on G/H.

Theorem 6.6 *Let H be a subgroup of a group G. If the operation on G/H given by $(g_1H) \cdot (g_2H) = (g_1g_2)H$ is well-defined, then it defines a group structure on G/H and $\pi : G \to G/H$ is a group homomorphism.*

Proof The verification of the group axioms is trivial. For instance, noticing that the coset eH is precisely the set H, we show that H is an identity element for the given operation. Let $gH \in G/H$ be an arbitrary element. Then

$$H \cdot gH = eH \cdot gH = (e \cdot g)H = gH$$

and similarly $gH \cdot H = gH$. We leave the verification of the other group axioms to the reader. The claim that π is a group homomorphism is nothing but the observation that $\pi(g_1g_2) = (g_1g_2)H = g_1H \cdot g_2H = \pi(g_1)\pi(g_2)$. □

We thus see that as soon as the operation $(g_1H) \cdot (g_2H) = (g_1g_2)H$ is well-defined, the quotient set G/H acquires a group structure. For the operation to be well-defined, one needs to verify that if $g_1H = g_3H$ and $g_2H = g_4H$, then necessarily $g_1g_2H = g_3g_4H$. In other words, given that $g_1^{-1}g_3 \in H$ and that $g_2^{-1}g_4 \in H$, it must follow that $(g_1g_2)^{-1}g_3g_4 \in H$. Let us denote $g_1^{-1}g_3 = h_1$ and $g_2^{-1}g_4 = h_2$. Noting that

$$(g_1g_2)^{-1}g_3g_4 = g_2^{-1}g_1^{-1}g_3g_4 = g_2^{-1}h_1g_4,$$

if G were abelian, then we would obtain that

$$(g_1g_2)^{-1}g_3g_4 = g_2^{-1}g_4h_1 = h_2h_1 \in H,$$

as desired. However, G need not be abelian. A closer look though reveals that a commutativity demand is far too strong. Indeed, if H is a normal subgroup of G, then for g_4 and h_1 as above, there exists $h' \in H$ such that $h_1g_4 = g_4h'$, and then

$$(g_1g_2)^{-1}g_3g_4 = g_2^{-1}h_1g_4 = g_2^{-1}g_4h' = h_1h' \in H.$$

We summarize this discussion as follows, remarking first that for a normal subgroup H in G, it follows at once that the left and right cosets of H in G coincide, and thus one may simply speak of the cosets of H in G.

Theorem 6.7 *If H is a normal subgroup of G, then the quotient set G/H of all cosets of H in G is a group when endowed with the operation*

$$(g_1H) \cdot (g_2H) = (g_1g_2)H.$$

The canonical projection $\pi : G \to G/H$ is then a group homomorphism. The group G/H is called the quotient group *of G by H.*

Corollary 6.2 *A subgroup H of a group G is normal if, and only if, H is the kernel of some group homomorphism whose domain is G.*

Proof It was already noted that the verification that the kernel of a homomorphism is a normal subgroup is straightforward. For the converse, show that

$$\text{Ker}(\pi) = H$$

for the canonical projection $\pi : G \to G/H$. □

The situation now for a topological group G and a subgroup H of it is summarized in the following result whose proof is left for the reader.

Theorem 6.8 *Let G be a topological group and H a subgroup.*

1. The set $\{gH \mid g \in G\}$ of all left cosets of H in G is the quotient set for the equivalence relation $g_1 \sim g_2 \iff g_1^{-1}g_2 \in H$. When endowed with the quotient topology it is a topological space but it need not be a group.
2. The set $\{Hg \mid g \in G\}$ of all right cosets of H in G is the quotient set for the equivalence relation $g_1 \sim g_2 \iff g_1 g_2^{-1} \in H$. When endowed with the quotient topology it is a topological space but it need not be a group.
3. If H is normal in G, then left and right cosets coincide, as do the two quotient sets above, which are then denoted by G/H. The quotient set G/H acquires a group structure from G and with the quotient topology it is a topological group.

6.5 Uniformities

In the presence of a topology notions such as convergence and continuity become available. However, a topology is too weak to speak of Cauchy sequences (and thus of completeness) or of uniform continuity, notions that do exist in the presence of a metric. However, there exists a structure, known as a uniformity, which is in between a topology and a metric, and which does allow one to speak of Cauchy sequences, completeness, completions, and uniform continuity. Moreover, every topological group is automatically endowed with two (often distinct) such uniformities, as we show below.

We first define the concept of uniformity in general and then proceed to describe the canonical uniformities present in any topological group. As motivation, recall that the axiomatization of a topology can be seen as the result of purifying a distance function, eliminating any trace of details irrelevant for continuity. Similarly, the axiomatization of a uniformity can be seen to arise from a similar purification of a distance function from anything irrelevant to uniform continuity.

Definition 6.7 Let X be a set. A non-empty collection $\mathscr{E} = \{E_i\}_{i \in I}$ of relations $E_i \subseteq X \times X$ is said to be a *uniform structure* or a *uniformity* on X if the following conditions hold (where we write $x \sim_E y$ to indicate that $(x, y) \in E$).

1. $x \sim_E x$ for all $E \in \mathscr{E}$ and $x \in X$.
2. For all $E \in \mathscr{E}$ there exists an $F \in \mathscr{E}$ such that $x \sim_F y$ implies $y \sim_E x$.
3. For all $E \in \mathscr{E}$ there exists an $F \in \mathscr{E}$ such that $x \sim_F y$ and $y \sim_F z$ together imply that $x \sim_E y$.
4. If $E, F \in \mathscr{E}$, then so is $E \cap F$.
5. If $E \subseteq F \subseteq X \times X$ and $E \in \mathscr{E}$, then $F \in \mathscr{E}$.

The pair $(X, \mathscr{E})$ is then called a *uniform space*. The members of $\mathscr{E}$ are knows as *entourages*. If context clarifies any ambiguity, it is common to refer to a uniform space X, leaving the collection $\mathscr{E}$ of entourages implicit.

Example 6.5 Given a metric space (X, d), a uniformity on the set X is induced by the distance function as follows. For every $\varepsilon > 0$ consider the set

$$E_\varepsilon = \{(x, y) \in X \times X \mid d(x, y) < \varepsilon\}.$$

It then follows that the collection

$$\mathscr{E} = \{E \subseteq X \times X \mid \exists \varepsilon > 0 \ E \supseteq E_\varepsilon\}$$

is a uniformity. The verification of the axioms is straightforward. For instance, condition 3 follows from the triangle inequality: given $E \supseteq E_\varepsilon$, let $F = E_{\varepsilon/2}$, then if $x \sim_F y$ and $y \sim_F z$, namely $d(x, y), d(y, z) < \varepsilon/2$, then $d(x, z) < \varepsilon$, so that $x \sim_E z$. The resulting uniformity is called the *induced uniformity*.

Definition 6.8 A function $f : X \to Y$ between uniform spaces is said to be *uniformly continuous* if $f^{-1}(E)$ is an entourage in X for every entourage E in Y (here $f^{-1}(E)$ stands for $\{(x, x') \in X \times X \mid f(x) \sim_E f(x')\}$).

Recall that a topology captures the notion of continuity in the sense that a function $f : X \to Y$ between metric spaces is continuous in the metric sense if, and only if, f is continuous with respect to the induced topologies. We now establish the analogous result for uniformities and uniformly continuous functions.

Theorem 6.9 *A function $f : X \to Y$ between metric spaces is uniformly continuous in the metric sense if, and only if, f is uniformly continuous with respect to the induced uniformities.*

Proof Suppose that f is uniformly continuous in the sense that for every entourage F in Y, $f^{-1}(F)$ is an entourage in X. Given $\varepsilon > 0$ consider the entourage

$$F_\varepsilon = \{(y, y') \in Y \times Y \mid d(y, y') < \varepsilon\}.$$

By definition of the induced uniformity, $f^{-1}(F_\varepsilon)$ contains an entourage of the form $E_\delta = \{(x, x') \in X \times X \mid d(x, x') < \delta\}$, for some $\delta > 0$. In particular, if $d(x, x') < \delta$, then $(x, x') \in E_\delta$ and thus $(f(x), f(x')) \in F_\varepsilon$, namely $d(f(x), f(x')) < \varepsilon$. In other words, f is uniformly continuous with respect to the metrics.

In the other direction, suppose that f is uniformly continuous in the sense that for all $\varepsilon > 0$ there exists a corresponding $\delta > 0$ such that $d(f(x), f(x')) < \varepsilon$ provided that $d(x, x') < \delta$. To show that f is uniformly continuous with respect to the induced uniformities, let F be an entourage in Y, namely there exists an $\varepsilon > 0$ such that $F \supseteq F_\varepsilon$ where

$$F_\varepsilon = \{(y, y') \in Y \times Y \mid d(y, y') < \varepsilon\}.$$

With $\delta > 0$ corresponding to ε, we claim that $f^{-1}(F) \supseteq E_\delta$, where

$$E_\delta = \{(x, x') \in X \times X \mid d(x, x') < \delta\}.$$

Indeed, if $(x, x') \in E_\delta$, then $d(x, x') < \delta$ and thus $d(f(x), f(x')) < \varepsilon$. It follows that $(x, x') \in f^{-1}(F_\varepsilon)$, and thus $E_\delta \subseteq f^{-1}(F_\varepsilon) \subseteq f^{-1}(F)$, This shows that $f^{-1}(E)$ contains an entourage in X, and thus is itself an entourage, which is what we needed to show. □

Clearly, the distance function can not be reconstructed from the induced uniformity. It is a simple matter to find different metrics that induce the same uniformity (for instance, by scaling all distances by a positive constant). A uniform space $(X, \mathscr{E})$ whose uniform structure $\mathscr{E}$ is the induced uniformity for some distance function d on X is said to be a *metrizable* uniform space. Thus, so far, we established that the concept uniform space is weaker than that of metric space, and that it correctly captures the notion of uniformly continuous functions. Next, we show that every uniform space induces a topology.

Theorem 6.10 *Let $(X, \mathscr{E})$ be a uniform space. The collection τ of sets $U \subseteq X$ such that for all $x \in U$ there exists an entourage $E \in \mathscr{E}$ such that $\{y \in X \mid x \sim_E y\} \subseteq U$ is a topology on U, called the* induced topology.

Proof The proof is left as an exercise for the reader. □

The passage from a distance function to the induced uniformity loses information and is thus not a reversible process. Similarly the passage from a uniformity to the induced topology is not reversible. A topological space (X, τ) whose topology τ is the induced topology for some uniformity on X is called a *uniformizable* topological space.

We are now ready to describe the two uniformities present on any topological group. Recall that every element $a \in G$ gives rise to the left translation map $g \mapsto ag$ and to the right translation map $g \mapsto ga$, both of which are homeomorphisms. We extend the translation mappings so that they also operate on subsets $S \subseteq G$, that is we define $aS = \{as \mid s \in S\}$, and similarly $Sa = \{sa \mid s \in S\}$.

Theorem 6.11 *Let G be a topological group and $\mathscr{B}$ a local basis at e. For every open set $U \in \mathscr{B}$ let $R_U = \{(x, y) \in X \times X \mid xy^{-1} \in U\}$. The collection*

$$\mathscr{E}_R = \{E \subseteq X \times X \mid \exists U \in \mathscr{B} \quad E \supseteq R_U\}$$

is a uniformity on X called the induced right uniformity. *Similarly, for each $U \in \mathscr{B}$ let $L_U = \{(x, y) \in X \times X \mid x^{-1}y \in U\}$. The collection*

$$\mathscr{E}_L = \{E \subseteq X \times X \mid \exists U \in \mathscr{B} \quad E \supseteq L_U\}$$

is a uniformity on X called the induced left uniformity.

Proof This is a routine verification of the axioms of a uniform space. □

The following properties of the induced uniformities justify the names given to them.

Theorem 6.12 *Let G be a topological group and let $\mathscr{E}_L$ and $\mathscr{E}_R$ be the left and right induced uniformities. Then*

1. *The induced topology by either $\mathscr{E}_L$ or $\mathscr{E}_R$ is the original topology on G.*
2. *Every left translation $g \mapsto ag$ is uniformly continuous with respect to $\mathscr{E}_L$.*
3. *Every right translation $g \mapsto ga$ is uniformly continuous with respect to $\mathscr{E}_R$.*

Proof The proof is straightforward. □

Remark 6.4 We conclude the book with the following remark. A sequence $\{x_m\}_{m\geq 1}$ in a uniform space X is said to be a *Cauchy* sequence if for every entourage E there exists an $N \in \mathbb{N}$ such that $x_k \sim_E x_n$ for all $k, n > N$. A uniform space X is said to be *complete* if every Cauchy sequence in X converges (under the topology induced by the uniformity). The process of completion of a metric space (as given in Chap. 4 by means of Cauchy sequences) extends to a completion process for every uniform space (however not by means of Cauchy sequences, as they are too weak for general uniform spaces). Similarly, other uniform concepts of metric spaces have analogous results valid for all uniform spaces.

The result above shows that the topology of a topological group is always uniformizable (and the uniformity can be chosen to further be compatible with all right or all left translations). Thus, in a sense, the topology of topological groups is rather tame, behaving more like metric spaces do than the more wild topological spaces out there do.

Exercises

Exercise 6.1 Let G be a group and $\text{Aut}(G)$ the set of all isomorphisms $\psi : G \to G$. Prove that, under composition of functions, $\text{Aut}(G)$ is a group. (This is the group of symmetries of the group G).

Exercise 6.2 Given a group G and an element $g \in G$, prove that $h \mapsto ghg^{-1}$ is an isomorphism $G \to G$. Such an isomorphism is called an *inner* isomorphism. Prove that the set of all inner isomorphisms of G is a normal subgroup of $\text{Aut}(G)$.

Exercise 6.3 Prove that if H is a normal subgroup of G, then $gH = Hg$ for all $g \in G$. Use a suitable subgroup H of the dihedral group D_3, the group of symmetries of an equilateral triangle, to refute the converse.

Exercise 6.4 Prove that if G is a topological group and if, for some $g \in G$, the set $\{g\}$ is closed, then G is Hausdorff.

Exercise 6.5 Prove that for all topological groups G, the quotient group $G/\overline{\{e\}}$ exists and is a Hausdorff topological group.

Exercise 6.6 Prove that an open subgroup H of a topological group G is also closed. (Hint: The complement of H is a union of translates of H).

Exercise 6.7 Prove that the topology induced by the uniformity induced by a metric is identical to the topology induced by the metric.

Exercise 6.8 For subsets S and T of a topological group G, consider the point-wise product $ST = \{st \mid s \in S, t \in T\}$. Prove that if S and T are compact, then so is ST.

Exercise 6.9 Prove Theorems 6.10, 6.11, and 6.12.

Exercise 6.10 The following is a well-known theorem in the theory of uniform spaces: A uniform space $(X, \mathscr{E})$ is metrizable if, and only if, the uniformity $\mathscr{E}$ is generated by countably many entourages, where $\mathscr{E}$ is *generated* by the collection $\mathscr{F} \subseteq \mathscr{E}$ if

$$\mathscr{E} = \{E \subseteq X \times X \mid \exists F \in \mathscr{F} \quad E \supseteq F\}.$$

Prove that a first countable Hausdorff topological group is metrizable.

Further Reading

For a broad physics-oriented introduction to groups see [4] or, for an introduction to groups in the context of Quantum Mechanics, see [5]. For a friendly introduction to topological groups see Chap. 3 of [1] (and the rest of the book for a delightful deeper study of the subject). For an elementary introduction to uniform spaces alongside topological spaces see [2]. Finally, for a comprehensive introduction to topological groups, starting with an independent treatment of topological spaces and metric spaces, see [3].

References

1. J. Diestel, A. Spalsbury, *The Joys of Haar Measure*, Graduate Studies in Mathematics (American Mathematical Society, Providence, 2014)
2. I.M. James, *Topological and Uniform Spaces*, Undergraduate Texts in Mathematics (Springer, New York, 1987)
3. N.G. Markley, *Topological Groups: An Introduction* (Wiley, New York, 2010)
4. S. Sternberg, *Group Theory and Physics* (Cambridge University Press, Cambridge, 1995)
5. M. Tinkham, *Group Theory and Quantum Mechanics* (Dover Publications, Mineola, 2003)

Chapter 7
Solved Problems

7.1 Linear Spaces

7.1 Consider the linear space $C(\mathbb{R}, \mathbb{R})$ and the vectors $x(t) = \sin(t)$, $y(t) = \cos(t)$, and $z(t) = 1$. Show that x, y, z are linearly independent, while x^2, y^2, z^2 are not.

7.2 Let V be a linear space and $S \subseteq V$ a set of vectors. Prove that S is linearly independent if, and only if, every finite subset of S is linearly independent.

7.3 Consider $\mathbb{R}$ as a linear space over itself. Prove that $\mathbb{R}$ has precisely two linear subspaces. Now consider $\mathbb{R}$ as a linear space over $\mathbb{Q}$ and prove that it has infinitely many linear subspaces.

7.4 Let V and W be linear spaces over the field K, and $X \subseteq V$ an arbitrary subset of V. A *linear relation* in X is any expression

$$0 = \sum_{k=1}^{m} \alpha_k x_k,$$

where $m > 0$, $\alpha_1, \ldots, \alpha_m \in K$, and $x_1, \ldots, x_m \in X$. Let S be the span of X and consider an arbitrary function $f : X \to W$. Prove that f extends to a linear operator $F : S \to W$ if, and only if, for any linear relation in X as above, one has

$$0 = \sum_{k=1}^{m} \alpha_k f(x_k).$$

© Springer International Publishing Switzerland 2015

C. Alabiso and I. Weiss, *A Primer on Hilbert Space Theory*,
UNITEXT for Physics, DOI 10.1007/978-3-319-03713-4_7

Conclude that if $\mathscr{B}$ is a basis for V, then any function $f : \mathscr{B} \to W$ extends to a linear operator $F : V \to W$.

7.5 Let $T : V \to W$ be a linear isomorphism between linear spaces. Prove that T maps a basis for V to a basis for W.

7.6 Prove that every infinite dimensional linear space over $\mathbb{R}$ contains an isomorphic copy of $\mathbb{R}^n$, for all $n \geq 0$. Is there a linear space that contains an isomorphic copy of every linear space over $\mathbb{R}$?

7.7 Prove that the space M of 3×3 matrices with entries in $\mathbb{R}$ is spanned by the subspace of matrices of rank 2.

7.8 Compute the dimension of the linear space P_n, $n \geq 0$, of polynomials with real coefficients and degree at most n, and of the linear space P of all polynomials with real coefficients.

7.9 Let $A, B : V \to V$ be two invertible linear operators from a linear space to itself. Prove that if A and B commute, i.e., $AB = BA$, then so do A^{-1} and B^{-1}.

7.10 Prove that a linear space V over an arbitrary field K (if you like you can take $K = \mathbb{R}$ or $K = \mathbb{C}$) is infinite dimensional if, and only if, V is isomorphic to a proper linear subspace of itself.

7.2 Topological Spaces

7.11 (Separation properties) A topological space X is said to satisfy:

- the T_0 *separation axiom* if for all distinct points $x, y \in X$ there exists an open set U such that either $x \in U$ and $y \notin U$, or $y \in U$ and $x \notin U$;
- the T_1 *separation axiom* if for all distinct points $x, y \in X$ there exist two open sets U and V such that $x \in U$ and $y \notin U$, and $y \in V$ and $x \notin V$;
- the T_2 *separation axiom* if X satisfies the Hausdorff separations property, that is for all distinct points $x, y \in X$ there exist two open sets U and V such that $x \in U$, $y \in V$, and $U \cap V = \emptyset$.

Clearly, the separation properties are increasing in strength. For every separation property, give an example of a topological space satisfying it, but not the next one.

7.12 (Locally connected spaces) A topological space X is called *locally (path) connected* at x if, for every open set V with $x \in V$, there exists a (path) connected open set U with $x \in U \subseteq V$. The space X is said to be *locally (path) connected* if it is locally (path) connected at every $x \in X$.

Show that a locally (path) connected space need not be (path) connected, nor does a (path) connected space need be locally (path) connected. For the latter, consider the *comb space*, the subset C of $\mathbb{R}^2$ given by

$$C = \{(x, 0) \mid 0 \leq x \leq 1\} \cup \{(0, y) \mid 0 \leq y \leq 1\} \cup A_1 \cup A_2 \cup \cdots \cup A_n \cup \cdots$$

where $A_n = \{(1/n, t) \mid 0 \leq t \leq 1\}$. The set C is given the subspace topology of the Euclidean topology on $\mathbb{R}^2$. (Hint: Draw a picture of this space!)

7.13 (Locally compact spaces) A topological space X is called *locally compact* at $x \in X$ if there exists a compact set C and an open set U such that $x \in U \subseteq C$. The space X is *locally compact* when it is locally compact at every $x \in X$.

Show that every compact space is locally compact, but not every locally compact space is compact. Provide an example of a space that is not locally compact.

7.14 Consider

$$X = \left\{ A \in \mathrm{M}_2(\mathbb{R}) \mid A^2 = 0 \right\}.$$

Endow X with the subspace topology of $\mathrm{M}_2(\mathbb{R})$, where $\mathrm{M}_2(\mathbb{R})$ is endowed with the topology induced by the identification $\Phi\colon \mathrm{M}_2(\mathbb{R}) \xrightarrow{\cong} \mathbb{R}^4$ given by

$$\begin{pmatrix} a & b \\ c & d \end{pmatrix} \overset{\Phi}{\mapsto} (a, b, c, d),$$

and $\mathbb{R}^4$ is endowed with the Euclidean topology. Decide whether X is closed in $M_2(\mathbb{R})$, whether X is compact, and whether X is connected.

7.15 Consider the topological spaces

$$X_1 = \left\{ (x, 1) \in \mathbb{R}^2 \mid x \in \mathbb{R} \right\} \quad \text{and} \quad X_2 = \left\{ (x, 2) \in \mathbb{R}^2 \mid x \in \mathbb{R} \right\}$$

endowed with the subspace topology from the Euclidean topology on $\mathbb{R}^2$. On the space

$$X = X_1 \sqcup X_2,$$

the coproduct of X_1 and X_2, consider the equivalence relation

$$(x, 1) \sim (x', 2) \overset{\text{def}}{\Leftrightarrow} x = x' \neq 0.$$

Let $\tilde{X} = X/\sim$ be the quotient space. Decide whether:

1. $\tilde{X}$ is T_0, T_1, T_2;
2. $\tilde{X}$ is compact;
3. $\tilde{X}$ is path-connected.

7.16 Consider the spaces

$$X_1 = \mathbb{R}^2,$$

$$X_2 = \left\{(x, y, z) \in \mathbb{R}^3 \mid x^2 + y^2 + z^2 = 1\right\} - \left\{\begin{pmatrix}0\\0\\1\end{pmatrix}\right\},$$

$$X_3 = \left\{(y_1, y_2) \in \mathbb{R}^2 \mid y_1^2 + y_2^2 = 1\right\},$$

$$X_4 = \left\{(y_1, y_2) \in \mathbb{R}^2 \mid y_1^2 + y_2^2 + 2y_1 = 0\right\} \cup \left\{(y_1, y_2) \in \mathbb{R}^2 \mid y_1^2 + y_2^2 - 2y_1 = 0\right\}.$$

For every pair of spaces, decide whether the spaces are homeomorphic or not.

7.17 Construct a topological space where the closed sets are stable under countable unions, but not under all arbitrary unions.

7.18 Let $A \in \mathrm{M}_n(\mathbb{R})$ be a symmetric matrix whose eigenvalues are positive. Decide whether

$$W = \left\{x \in \mathbb{R}^n \mid x^{\mathrm{t}} A x = 1\right\}$$

is a compact subset of $\mathbb{R}^n$ with the Euclidean topology.

7.19 Consider a topological Hausdorff space X and a subset $D \subseteq X$ which is dense and locally compact. Is D necessarily open in X?

7.20 Consider the ring $\mathbb{Z}$ of integer numbers. The *spectrum* of $\mathbb{Z}$ is the set

$$X = \{p \cdot \mathbb{Z} \mid p \text{ prime or } p = 0\},$$

where $p \cdot \mathbb{Z} = \{p \cdot k \mid k \in \mathbb{Z}\}$. For all $a \in \mathbb{Z}$ define the set

$$V(a \cdot \mathbb{Z}) = \{p \cdot \mathbb{Z} \in X \mid a \cdot \mathbb{Z} \subseteq p \cdot \mathbb{Z}\}.$$

- Prove that the collection $\mathscr{V} = \{V(a \cdot \mathbb{Z}) \mid a \in \mathbb{Z}\}$ is the set of closed subsets of a topology on X. That topology is known as the *Zarisky topology* on $\mathbb{Z}$.
- Does there exist a single point in X whose closure is the whole space?
- Is X Hausdorff?

7.3 Metric Spaces

7.21 Let (X, d) be a metric space. Suppose that there exists an infinite countable set $\{x_n\}_{n \in \mathbb{N}}$ of points such that $d(x_k, x_m) = 1$ for all $k, m \in \mathbb{N}$ with $k \neq m$. Prove that X is not compact.

7.22 Consider the set ℓ_∞ of all bounded sequences of real numbers with the metric induced by the ℓ_∞ norm, i.e.,

$$d(x, y) = \sum_n |x_n - y_n|.$$

Prove that ℓ_∞ is connected and not compact.

7.23 Let $f : \mathbb{R} \to \mathbb{R}$ be an infinitely differentiable function with the property that for all $x_0 \in \mathbb{R}$ there exists a natural number $n \geq 0$ with $f^{(n)}(x_0) = 0$. Prove that there exists an open interval $I = (a, b)$, with $a < b$, such that f agrees with a polynomial function on I.

7.24 Consider the space $C([0, 1], \mathbb{R})$ of all continuous real-valued functions on the interval $[0, 1]$, endowed with the L_∞ norm

$$\|f\|_\infty = \max_{t\in[0,1]} |f(t)|.$$

Is this norm induced by any inner product on $C([0, 1], \mathbb{R})$?

7.25 Prove that the empty set and every singleton set admit a unique metric, while every other set admits infinitely many non-isometric metric structures.

7.26 Let (X, d) be a metric space. Consider a subset S of X. Prove that the closure $\overline{S}$ of S in X is the set

$$\{x \in X \mid d(x, S) = 0\},$$

where $d(x, S) = \inf_{s\in S} d(x, s)$.

7.27 Let (X, d) be a metric space. Prove that the function $\tilde{d}\colon X \times X \to \mathbb{R}$ defined by $\tilde{d}(x, y) = \min\{1, d(x, y)\}$ is a metric on X which induces the same topology as d does.

7.28 Consider the spaces $\mathbb{R}$ and $\mathbb{R}^2$ endowed with the Euclidean metrics. Decide whether there exists an isometry $f\colon \mathbb{R}^2 \to \mathbb{R}$.

7.29 Let $\{d_i\}_{i\in I}$ be a family of distance functions on a given set X. Prove that the supremum $d_s : X \times X \to \mathbb{R}_+$ given by

$$d_s(x, y) = \sup_{i\in I}\{d_i(x, y)\}$$

is a distance function on X. In contrast, show that the similarly defined infimum of distance functions need not be a distance function.

7.30 Let $f : X \to Y$ be a function between metric spaces.

1. Is continuity, or uniform continuity, a sufficient condition to ensure that if X is complete then $f(X)$ is complete?
2. Is continuity, or uniform continuity, a sufficient conditon to ensure that if X is totally bounded then $f(X)$ is totally bounded?
3. Is continuity, or uniform continuity, a sufficient condition to ensure that if X is complete and totally bounded then $f(X)$ is complete and totally bounded?

7.4 Normed Spaces and Banach Spaces

7.31 A metric function d on a linear space V, with ground field $K = \mathbb{R}$ or $K = \mathbb{C}$, is said to be *translation invariant* when

$$d(x, y) = d(x + z, y + z)$$

for all $x, y, z \in V$ and is said to be *scale homogenous* when

$$d(\alpha x, \alpha y) = |\alpha| d(x, y)$$

for all $x, y \in V$ and $\alpha \in K$. Prove that V is a normed space if, and only if, it is endowed with a translation invariant and scale homogenous metric.

7.32 Let $X = C([a, b], \mathbb{R})$ be the space of continuous functions $f : [a, b] \to \mathbb{R}$ and consider the assignment

$$f \mapsto \|f\| = \int_a^b |f(x)|\, dx$$

for all $f \in X$.

1. Give a direct proof that $f \mapsto \|f\|$ is a norm on X.
2. Decide whether X with the induced metric $d(f, g) = \|f - g\|$ is a complete metric space.

7.33 Consider $\mathbb{R}$ with the function $d : \mathbb{R} \times \mathbb{R} \to \mathbb{R}_+$ given by

$$d(x, y) = \frac{|x - y|}{1 + |x - y|}$$

for all $x, y \in \mathbb{R}$. Decide whether d is a distance function on $\mathbb{R}$ and if so whether the metric is induced by a norm.

7.34 Let X be a linear space endowed with two norms, $\|\cdot\|_1$ and $\|\cdot\|_2$. Suppose that there exists $K > 0$ such that

$$\|v\|_1 \le K \cdot \|v\|_2$$

holds for all $v \in X$. Prove that the topology induced by $\|\cdot\|_2$ is finer than the topology induced by $\|\cdot\|_1$.

7.35 For vectors x, y in a linear space V, let $L(x, y)$ be the *affine line* connecting x and y, that is

$$L(x, y) = \{\alpha x + (1 - \alpha)y \mid 0 \le \alpha \le 1\}.$$

A normed space V is said to be *strictly convex* if for all distinct vectors x, y on the unit sphere, i.e., $\|x\| = \|y\| = 1$, the line $L(x, y)$ intersects the unit sphere only at x and y.

1. Prove that for x, y with $\|x\| = \|y\| = 1$ the line $L(x, y)$ is fully contained in the unit ball $\{z \in V \mid \|z\| \le 1\}$.
2. Prove that a normed space induced by an inner product is strictly convex.
3. Prove that in a strictly convex normed space, if $x \ne y$ are vectors which satisfy $\|x\| = \|y\| = 1$, then $\|x + y\| < 2$.
4. Of the ℓ_1, ℓ_2, and ℓ_∞ norms on $\mathbb{R}^2$, decide which, if any, are strictly convex.

7.36 Prove that the kernel of a bounded linear operator $A : V \to W$ is a closed linear subspace of V.

7.37 Consider the normed space ℓ_2 and the linear operator $A : \ell_2 \to \ell_2$ given by

$$A(\{x_n\}_{n\ge1}) = \{(1 - \frac{1}{n})x_n\}.$$

Prove that A is a bounded linear operator that never attains its norm, i.e., there exists no x_0 such that

$$\|Ax_0\| = \|A\|\|x_0\|.$$

7.38 Let $A : U \to V$ and $B : V \to W$ be bounded linear operators between normed spaces. Prove that the composition BA is a bounded linear operator and that

$$\|BA\| \le \|B\|\|A\|.$$

7.39 Let $\mathscr{B}$ be a Banach space and recall that $\mathbf{B}(\mathscr{B})$, the set of all bounded linear operators $A : \mathscr{B} \to \mathscr{B}$, is a normed space when endowed with the operator norm, and thus is a topological space. Prove that the set $G \subseteq \mathbf{B}(\mathscr{B})$ consisting of the invertible operators is an open subset of $\mathbf{B}(\mathscr{B})$.

7.40 Let $\mathscr{B}_1, \mathscr{B}_2, \mathscr{B}_3$ be Banach spaces with $T : \mathscr{B}_1 \to \mathscr{B}_2$ a bounded linear operator (whose domain is all of $\mathscr{B}_1$) and let $S : \mathscr{B}_2 \to \mathscr{B}_3$ be a closed linear operator whose domain is $\mathscr{D}(S) \subseteq \mathscr{B}_2$. Prove that ST is a closed linear operator.

7.5 Topological Groups

7.41 Prove that the group $(\mathbb{R}^n, +)$ with the Euclidean topology is a Hausdorff topological group. In fact, it is a topological vector space, i.e., all of the linear space structure mappings are continuous.

7.42 Prove that the *special linear group*

$$\mathrm{SL}_n(\mathbb{R}) = \{A \in \mathrm{GL}_n(\mathbb{R}) \mid \det(A) = 1\}$$

is a topological group.

7.43 Prove that the *orthogonal group*

$$\mathrm{O}(n) = \left\{A \in \mathrm{GL}_n(\mathbb{R}) \mid A^{-1} = A^{\mathrm{t}}\right\}$$

is a topological group.

7.44 Prove that the *special orthogonal group*

$$\mathrm{SO}(n) = \{A \in \mathrm{O}(n) \mid \det(A) = 1\}$$

is a topological group.

7.45 Prove that the *symplectic group*

$$\mathrm{Sp}_{2n}(\mathbb{R}) = \left\{A \in \mathrm{GL}_{2n}(\mathbb{R}) \mid A^{\mathrm{t}} \cdot J_0 \cdot A = J_0\right\},$$

where

$$J_0 = \left(\begin{array}{c|c} 0 & \mathrm{id}_n \\ \hline -\mathrm{id}_n & 0 \end{array}\right),$$

is a topological group.

7.46 Prove that the topological group $\mathrm{SO}(n)$ is path-connected.

7.47 Prove that the topological group $\mathrm{O}(n)$ is the disjoint union of two path-connected components, $\mathrm{O}^+(n)$ and $\mathrm{O}^-(n)$.

7.48 Prove that

$$\mathrm{GL}_n^+(\mathbb{R}) = \{A \in \mathrm{GL}_n(\mathbb{R}) \mid \det(A) > 0\}$$

is a path-connected component of the topological group $\mathrm{GL}_n(\mathbb{R})$.

7.49 Prove that the topological group $\mathrm{GL}_n(\mathbb{R})$ has two connected components, more precisely,

$$\mathrm{GL}_n^+(\mathbb{R}) = \{A \in \mathrm{GL}_n(\mathbb{R}) \mid \det(A) > 0\}$$

and

$$\mathrm{GL}_n^-(\mathbb{R}) = \{A \in \mathrm{GL}_n(\mathbb{R}) \mid \det(A) < 0\}.$$

7.50 Prove that the special linear group

$$\mathrm{SL}_n(\mathbb{R}) = \{A \in \mathrm{GL}_n(\mathbb{R}) \mid \det(A) = 1\}$$

is path-connected.

7.6 Solutions

7.6.1 Linear Spaces

7.1 Suppose that

$$\alpha x + \beta y + \gamma z = 0$$

for some scalars $\alpha, \beta, \gamma \in \mathbb{R}$. The equation is a functional equality, and thus for any choice of $t \in \mathbb{R}$ we have that

$$\alpha x(t) + \beta y(t) + \gamma z(t) = 0.$$

However, by considering the values $t = 0, t = \pi/2, t = \pi$, we obtain the equations: $\beta + \gamma = 0$, $\alpha + \gamma = 0$, and $-\beta + \gamma = 0$, from which $\alpha = \beta = \gamma = 0$ follows easily. Thus the only linear combination resulting in the constantly zero function is the trivial combination, showing the desired linear independence. Now, as for x^2, y^2, and z^2, noting that $x^2(t) + y^2(t) = 1 = z^2(t)$, we see that $x^2 + y^2 - z^2 = 0$, establishing the desired linear dependence.

7.2 Given a set S of vectors in V, suppose first that S is linearly independent. We must show that every finite subset of it is linearly independent, thus let $S_f \subseteq S$ be a finite subset of S. Suppose that

$$0 = \sum_{k=1}^{m} \alpha_k x_k,$$

where $\alpha_1, \ldots, \alpha_m$ are scalars and $x_1, \ldots, x_m \in S_f$. But then the exact same linear combination is also a linear combination of elements from the linearly independent set S, and thus all scalars must be 0. This shows S_f is linearly independent. In the other direction, suppose every finite subset of S is linearly independent. To show that S is linearly independent suppose that 0 is obtained as a linear combination (necessarily a finite one!) as above, with $x_1 \ldots, x_m \in S$. Consider then the finite set

$$\{x_1, \ldots, x_m\} \subseteq S$$

and we thus see that 0 is written as a linear combination of elements from that subset, which is linearly independent by assumption. Thus all scalars must be 0, and thus S is linearly independent.

7.3 Consider $\mathbb{R}$ as a linear space over itself and let W be a non-trivial linear subspace of $\mathbb{R}$. Thus W contains some non-zero vector, namely a real number $x \neq 0$. The dimension of $\mathbb{R}$ over itself is clearly 1, and thus any non-zero vector spans all of $\mathbb{R}$. Since W, as a linear subspace, is its own span, we conclude that $W = \mathbb{R}$. Thus, we showed that other than the trivial subspace $\{0\}$, the only other subspace of $\mathbb{R}$ is $\mathbb{R}$ itself, as required.

Now we consider $\mathbb{R}$ as a linear space over $\mathbb{Q}$, and note that this is an infinite dimensional linear space. Let $\mathscr{B}$ be a Hamel basis for $\mathbb{R}$ over $\mathbb{Q}$. For every subset X of $\mathscr{B}$ consider the span of X, and denote it by W_X. We claim that $W_X \neq W_Y$ for all subsets $X, Y \subseteq \mathscr{B}$ with $X \neq Y$. Indeed, seeking a contradiction, suppose $W_X = W_Y$ and we may assume that X is not a subset of Y. Choose a vector $x \in X - Y$. Since $x \in W_X$ it follows that $x \in W_Y$. Thus

$$x = \sum_{k=1}^{m} \alpha_k y_k,$$

where $\alpha_1, \ldots, \alpha_m \in \mathbb{Q}$ and $y_1, \ldots, y_m \in Y$. But then

$$0 = x - \sum_{k=1}^{m} \alpha_k y_k$$

is a non-trivial linear combination of vectors from the linearly independent set $\mathscr{B}$, an impossibility. To conclude then, for each subset of $\mathscr{B}$ we obtained a unique linear subspace of $\mathbb{R}$. There are thus infinitely many subspaces, as required. In fact, the cardinality of a Hamel basis in this case was shown to be equal to $|\mathbb{R}|$, the cardinality of the real numbers. We thus showed that $\mathbb{R}$ as a linear space over $\mathbb{Q}$ has at least $|\mathscr{P}(\mathscr{B})|$ many linear subspaces, a cardinality known to be strictly larger than the cardinality of $\mathscr{B}$, and thus of $\mathbb{R}$.

7.4 Suppose first that the given function $f : X \to W$ extends to a linear operator $F : S \to W$. If

$$0 = \sum_{k=1}^{m} \alpha_k x_k,$$

is a linear relation in X, then

$$0 = F(0) = F(\sum_{k=1}^{m} \alpha_k x_k) = \sum_{k=1}^{m} \alpha_k F(x_k) = \sum_{k=1}^{m} \alpha_k f(x_k),$$

as claimed. In the other direction, suppose the condition on linear relations in X is met. Given $x \in S$, write

$$y = \sum \alpha_x \cdot x$$

as a (finite!) linear combination of vectors from X. If

$$F(y) = \sum \alpha_x \cdot f(x)$$

is well-defined, then it clearly gives rise to the desired linear extension of f. So, suppose that

$$y = \sum \beta_x \cdot x$$

is another expression of y as a linear combination of vectors from X. But then, by subtracting the two expressions, we obtain

$$0 = \sum (\alpha_x - \beta_x) \cdot x$$

which is a linear relation in X. It thus follows that

$$0 = \sum (\alpha_x - \beta_x) \cdot f(x)$$

from which it follows that the formula for computing $F(y)$ is independent of the presentation of y as a linear combination of vectors from X, as desired.

As for the special case where $X = \mathscr{B}$ is a basis, note that the span of X is then the entire ambient space V, and that the linear independence of $\mathscr{B}$ implies that there are no linear relations in X, so the needed condition for the extension is vacuously satisfied.

7.5 Suppose that $T : V \to W$ is an isomorphism and that $\mathscr{B}$ is a basis for V. We will show that $T(\mathscr{B}) = \{T(b) \mid b \in \mathscr{B}\}$ is a basis for W. Given an arbitrary $w \in W$, consider $T^{-1}(w) \in V$. We may write $T^{-1}(w)$ as a linear combination

$$T^{-1}(w) = \sum \alpha_b \cdot b$$

and thus

$$w = T(T^{-1}(w)) = \sum \alpha_b \cdot T(b)$$

showing that $T(\mathscr{B})$ spans W. Next, suppose a linear combination

$$0 = \sum \alpha_b \cdot T(b)$$

is given. But

$$\sum \alpha_b \cdot T(b) = T(\sum \alpha_b \cdot b)$$

and thus $\sum \alpha_b \cdot b \in \mathrm{Ker(T)} = \{0\}$ (since T is injective). It follows that

$$\sum \alpha_b \cdot b = 0$$

and therefore that all of the coefficients are 0, as required.

7.6 Suppose V is an infinite dimensional linear space and let $\mathscr{B}$ be a basis for V. Choose a countable subset $\{y_1, y_2, \ldots\} \subseteq \mathscr{B}$ and let W_n be the span of $\{y_1, \ldots, y_n\}$. It is immediate to verify that $T : \mathbb{R}^n \to V$ given by

$$T(x) = T(x_1, \ldots, x_n) = \sum_{k=1}^{n} x_k y_k$$

is a linear isomorphism identifying a copy of $\mathbb{R}^n$ inside V, as claimed.

As for the existence of a linear space $\mathbf{U}$ which contains an isomorphic copy of every linear space over $\mathbb{R}$, the fact that no such linear space exists is a consequence of a set-theoretic result known as Cantor's Theorem. Suppose that such a $\mathbf{U}$ does exist. By constructing free linear spaces one sees that linear spaces of arbitrarily large cardinality exist. That is, given any set X, no matter how large, there always exists a linear space having X as a basis. Since $\mathbf{U}$ is assumed to contain a copy of every linear space, it follows that X is in bijection with a subset of $\mathbf{U}$. In other words, there exists an injection $X \to \mathbf{U}$, and thus $|X| \leq |\mathbf{U}|$, for all sets X.

In particular, for the set $X = \mathscr{P}(\mathbf{U})$ of all subsets of $\mathbf{U}$ one has that

$$|\mathscr{P}(\mathbf{U})| \leq |\mathbf{U}|.$$

However, the function $u \mapsto \{u\}$ is clearly an injection $\mathbf{U} \to \mathscr{P}(\mathbf{U})$, and thus

$$|\mathbf{U}| \leq |\mathscr{P}(\mathbf{U})|.$$

We thus conclude, by the Cantor-Shröder-Bernstein Theorem (see Preliminaries if needed), that

$$|\mathbf{U}| = |\mathscr{P}(\mathbf{U})|.$$

However, this is known to be impossible by an argument we now present.

Theorem 7.1 (Cantor's Theorem) *For all sets S there exists no surjective function $S \to \mathscr{P}(S)$. In particular $|S| < |\mathscr{P}(S)|$.*

Proof Suppose a surjective function $f : S \to \mathscr{P}(S)$ exists. Consider then the set

$$S_! = \{s \in S \mid s \notin f(s)\}$$

which is clearly a subset of S, and thus $S_! \in \mathscr{P}(S)$. Since f is surjective, it follows that there exists $s_! \in S$ with $f(s_!) = S_!$. To obtain a contradiction let us consider whether $s_! \in S_!$ or not. If $s_! \in S_!$, then $s_! \notin f(s_!) = S_!$, which is absurd. However, if $s_! \notin S_! = f(s_!)$, then $s_!$ fulfills the condition for being an element in $S_!$, and thus $s_! \in S_!$, again an absurdity. We must conclude that no such f exists.

Since no surjection from S to $\mathscr{P}(S)$ exists, no bijection exists either, and thus the sets have different cardinalities. Since $s \mapsto \{s\}$ is clearly an injection $S \to \mathscr{P}(S)$ it follows that $|S| < |\mathscr{P}(S)|$. □

7.7 Let E_{ab} be the 3×3 matrix whose entries are all 0 except for the (a, b) entry being 1. Clearly, the set $\{E_{ab}\}_{a,b\in\{1,2,3\}}$ spans M. We have to find a basis composed of rank 2 matrices. For $a, b, c, d \in \{1, 2, 3\}$, define

$$F_{a,b;c,d} = E_{ab} + E_{cd} \quad \text{and} \quad G_{a,b;c,d} = E_{ab} - E_{cd}.$$

Note that, if $a \neq c$ and $b \neq d$, then $F_{a,b;c,d}$ and $G_{a,b;c,d}$ have rank 2. Note also that

$$E_{ab} = \frac{1}{2}F_{a,b;c,d} + \frac{1}{2}G_{a,b;c,d},$$

where we can choose $a \neq c$ and $b \neq d$. The reader can now easily identify a set of rank 2 matrices that span M.

7.8 We prove that $\dim(P_n) = n + 1$ by showing that

$$\{1, x, x^2, \dots, x^n\}$$

is a basis. Clearly these vectors span P_n and the fact that they are linearly independent follows easily from the more general argument given next, where the dimension of P is computed.

We show that the dimension of P is countably infinite by showing that

$$\{1, x, x^2, \dots, x^n, \dots\}$$

is a basis. It is obvious that these vectors span P. To see that they are linearly independent suppose that a linear combination of the vectors results in the zero vector. In other words, $0 = \sum_{k=0}^{n} \alpha_k x^k$ for some real numbers $\alpha_0, \dots, \alpha_n$. But this linear combination is a polynomial and it is well-known that the only polynomial that evaluates to 0 on all real numbers is the 0 polynomial. Thus all coefficients must be equal to 0, and therefore the only linear combination resulting in the zero vector is the trivial linear combination.

7.9 We note first the general fact that if $A : U \to V$ and $B : V \to W$ are invertible functions, then $(B \circ A)^{-1} = A^{-1} \circ B^{-1}$. This is nothing but the computation

$$B(A(A^{-1}(B^{-1}(w)))) = B(B^{-1}(w)) = w$$

and similarly

$$A^{-1}(B^{-1}(B(A(u)))) = A^{-1}(A(u)) = u$$

for all $w \in W$ and all $u \in U$. Now, going back to the linear operators A and B, assumed to commute, we have

$$A^{-1}B^{-1} = (BA)^{-1} = (AB)^{-1} = B^{-1}A^{-1},$$

as required.

7.10 Assume first that V is infinite dimensional, and let $\mathscr{B}$ be an infinite basis of V. Fix a vector $x_0 \in \mathscr{B}$ and consider the set $S = \mathscr{B} - \{x_0\}$. As a subset of a basis these vectors are linearly independent, and thus S is a basis of the subspace spanned by S. Clearly, x_0 is missing from that span (otherwise the original $\mathscr{B}$ would be linearly dependent, which is impossible since it is a basis), and thus S is a proper subspace of V. To show that $S \cong V$ it suffices to show they have equal dimensions, namely that $|\mathscr{B}| = |S|$. Clearly, $|S| \le |\mathscr{B}|$, since the inclusion $S \to \mathscr{B}$ is an injection. If we can construct an injection in the other direction as well, then, using the Cantor-Shröder-Bernstein Theorem, the cardinalities will indeed be shown to be equal. To construct an injection $f : \mathscr{B} \to S$, let $\{x_m\}_{m \ge 1}$ be a countably infinite list of distinct vectors in S. Define now $f(x_m) = x_{m+1}$ for all $m \ge 1$, with $f(x_0) = x_1$, and $f(x) = x$ in all other cases. It is immediate to verify that f is indeed an injection. This completes the proof that if V is infinite dimensional, then it is isomorphic to a proper subspace of it.

In the other direction, suppose that $V \cong U$ for some proper subspace U. Since linear spaces over the same field are isomorphic if, and only if, their dimensions are equal, it follows that $\dim(V) = \dim(U)$. But in a finite dimensional space, if a subspace has the same dimension as the ambient space, then the subspace coincides with the ambient space, and is thus not proper. We conclude that V must be infinite dimensional.

7.6.2 Topological Spaces

Remark 7.1 In topology the term *neighborhood* of a point x in a topological space X may mean one of two things. Either an open set U in X with $x \in U$, or an arbitrary subset $N \subseteq X$ which contains an open set U with $x \in U$. The difference is only cosmetic, and one can easily translate between the two situations. In the solutions below we adhere to the first meaning of neighborhood.

7.11 Any indiscrete space with more than one element very clearly does not satisfy the T_0 property. For an example of a space that is T_0 but not T_1 consider the Sierpinski space $\mathbb{S} = \{0, 1\}$. For a T_1 space that is not T_2, let S be an infinite set endowed with the cofinite topology. Clearly the T_1 separation property is satisfied since given distinct points $x, y \in S$, the sets $S - \{x\}$ and $S - \{y\}$ are open and clearly satisfy the requirements. We show that S is not T_2. Indeed, let x, y be two distinct points in S and suppose $x \in U$ and y$\in V$ for some open sets U, V. Then the complements of U and of V are finite subsets of S, and thus

$$S - (U \cap V) = (S - U) \cup (S - V)$$

is also finite. Since S is infinite it follows that $U \cap V \neq \emptyset$.

7.12 A locally (path) connected space that is not (path) connected is, for instance, the space $(0, 1) \cup (3, 4)$ as a subspace of $\mathbb{R}$ with the Euclidean topology (as is easy to see). As for the comb space, we first show that it is path connected, and thus also connected. Indeed, any point admits a path to the point $(0, 0)$ by first traveling south to meet the X-axis and then traveling west. Thus any two points can be joined by a path in the space, so it is indeed path-connected. It is however not locally connected, and thus not locally path-connected. Indeed, given any point of the form $p = (0, y)$ with $0 < y < 1$, consider a small enough circle centered at p which does not intersect the X-axis. Any open set containing p must contain a subset of this form, but such a set is not connected.

7.13 Let X be a compact space. Given any $x \in X$ we may take $U = C = X$, and then $x \in U \subseteq C$ with U open and C compact, so X is locally compact. An example of a locally compact space that is not compact is, e.g., $\mathbb{R}$ with the Euclidean topology. Its non-compactness is obvious. As for it being locally compact, suppose $x \in \mathbb{R}$ is given. Then $x \in (x - 1, x + 1) \subseteq [x - 1, x + 1]$, as required.

As an example of a space that is not locally compact, consider $\mathbb{Q}$ as a subspace of $\mathbb{R}$ with the Euclidean topology. We will show that $\mathbb{Q}$ is not locally compact at any point. This will be done by showing that in fact no non-empty compact subset of $\mathbb{Q}$ contains a non-empty open set. To see that, suppose that $U \subseteq C$ where U is open and C is compact, and both are non-empty. It follows that $\mathbb{Q} \cap (a, b) \subseteq U$ for some non-empty open interval (a, b). Since $\mathbb{Q}$ is Hausdorff, the compact set C is closed, and thus, taking closures, we obtain that $[a, b] \cap \mathbb{Q} \subseteq C$. This is thus a closed subset of a compact set in a Hausdorff space, and thus $\mathbb{Q} \cap [a, b]$ is compact. But that is not the case. Indeed, let $y \in [a, b]$ be an irrational number and let $\{q_n\}_n \in \mathbb{N}$ be a strictly increasing sequence of rationals tending to y. Let $U_n = \mathbb{Q} \cap (-\infty, q_n)$ and consider also the set $\mathbb{Q} \cap (y, \infty)$. These are open sets in the subspace topology that cover $[a, b]$ but (as the reader may easily verify) they admit no finite subcovering.

7.14 We prove that X is closed. The condition

$$\begin{pmatrix} a & b \\ c & d \end{pmatrix} \in X$$

reads as

$$\begin{cases} a^2 + bc = 0 \\ ab + bd = 0 \\ bc + d^2 = 0 \\ ac + cd = 0. \end{cases}$$

In particular, by means of the homeomorphism $\Phi\colon \mathrm{M}_2(\mathbb{R}) \to \mathbb{R}^4$, the set X can be identified with the set

$$S = \{u = (a, b, c, d) \in \mathbb{R}^4 \mid \psi_1(u) = \psi_2(u) = \psi_3(u) = \psi_4(u) = 0\},$$

where

$$\psi_1, \psi_2, \psi_3, \psi_4\colon \mathbb{R}^4 \to \mathbb{R}$$

are the functions

$$\psi_1(x, y, z, w) = x^2 + yz, \quad \psi_2(x, y, z, w) = xy + yw,$$

$$\psi_3(x, y, z, w) = yz + z^2, \quad \psi_4(x, y, z, w) = xz + zw.$$

In other words,

$$S = \psi_1^{-1}(\{0\}) \cap \psi_2^{-1}(\{0\}) \cap \psi_3^{-1}(\{0\}) \cap \psi_4^{-1}(\{0\})$$

and since each ψ_i is a polynomial function, and thus continuous, the set $\phi_i^{-1}(\{0\})$ is closed (since $\{0\}$ is closed in $\mathbb{R}$). We see that S is the intersection of four closed sets, and hence it is itself closed.

We show now that X is not compact. Indeed, for all $n \in \mathbb{Z}$, consider the subset

$$U_n = \left\{ \begin{pmatrix} a & b \\ c & d \end{pmatrix} \in X \mid a \in (n-1, n+1),\, b \in \mathbb{R},\, c \in \mathbb{R},\, d \in \mathbb{R} \right\} \cap X.$$

Each U_n is an open set in X. Furthermore, one has that

$$\bigcup_{n \in \mathbb{Z}} U_n = X.$$

But there is no finite sub-family of $\{U_n\}_{n \in \mathbb{Z}}$ that still covers X. Indeed, for any $m \in \mathbb{Z}$, the matrix $\begin{pmatrix} m & 0 \\ 0 & 0 \end{pmatrix}$ belongs only to U_m.

Finally, we show that X is connected. In fact, we prove that X is path-connected. To do that, take any $A \in X$. We first notice that there exists a non-singular matrix $M \in M_2(\mathbb{R})$ such that

$$A = M^{-1} \cdot \begin{pmatrix} 0 & b \\ 0 & 0 \end{pmatrix} \cdot M,$$

for $b \in \mathbb{R}$. In other words, any matrix in X is similar to an upper-triangular matrix. Indeed, consider A as a matrix in $\mathrm{M}_2(\mathbb{C})$. Since $\mathbb{C}$ is algebraically closed, A can be triangulated in $\mathrm{M}_2(\mathbb{C})$, namely, it is similar to a matrix

$$\begin{pmatrix} a & b \\ 0 & c \end{pmatrix},$$

with a, b, c possibly complex numbers. The condition $A^2 = 0$ reads as

$$\begin{cases} a^2 = 0 \\ ab + bc = 0 \\ c^2 = 0, \end{cases}$$

hence the eigenvalues are $a = 0$ and $c = 0$. In particular, the eigenvalues belong to $\mathbb{R}$. Therefore A is triangulable also in $\mathrm{M}_2(\mathbb{R})$, proving the claim.

Now, consider the path

$$\gamma \colon [0, 1] \to X \subset \mathrm{M}_2(\mathbb{R}), \quad \gamma(t) = M^{-1} \cdot \begin{pmatrix} 0 & tb \\ 0 & 0 \end{pmatrix} \cdot M.$$

Note that, for any $t \in [0, 1]$, $(\gamma(t))^2 = 0$, and hence $\gamma(t) \in X$. Note also that $\gamma \colon [0, 1] \to X$ is continuous and that

$$\gamma(0) = \begin{pmatrix} 0 & 0 \\ 0 & 0 \end{pmatrix} \quad \text{and} \quad \gamma(1) = A,$$

proving that, for any $A \in X$, there exists a path connecting it to the zero matrix. This clearly suffices to show that any two elements in X can be connected by a path in X, and thus X is path-connected, as claimed.

7.15 1. We prove that $\tilde{X}$ is T_1 and hence also T_0. We have to prove that, for any pair of distinct points $p, q \in \tilde{X}$, there exists a neighborhood U_p of p not containing q and a neighborhood U_q of q not containing p. Consider first the following case. Suppose that the points are $p = [(x, 1)] = [(x, 2)]$ with $x \in \mathbb{R} - \{0\}$ and $q = \left[(x', y')\right]$ with $x' \in \mathbb{R}$ and $y' \in \{1, 2\}$. Take $r \in \mathbb{R}$ such that $0 < r < \left|x - x'\right|$. Then the sets $U_p = \left\{[(u, 1)] \in \tilde{X} \mid |u - x| < r\right\}$ and $U_q = \left\{\left[(u', y')\right] \in \tilde{X} \mid \left|u' - x'\right| < r\right\}$ are open neighborhoods of p, respectively q, and it holds that $p \notin U_q$ and $q \notin U_p$. It remains to consider just the case $p = [(0, 1)]$ and $q = [(0, 2)]$. In this case, we consider the sets $U_p = \left\{[(u, 1)] \in \tilde{X} \mid u \in \mathbb{R}\right\}$ and $U_q = \left\{[(v, 2)] \in \tilde{X} \mid v \in \mathbb{R}\right\}$. These are clearly neighborhoods of p, respectively q, and $p \notin U_q$ and $q \notin U_p$. The space $\tilde{X}$ is thus T_1.

We prove next that $\tilde{X}$ is not T_2. Consider the points $p = [(0, 1)]$, $q = [(0, 2)]$ for which we prove that no disjoint neighborhoods exist. Indeed, any neighborhood of p contains an open neighborhood of the type

$$\hat{U}_p = \left\{[(u, 1)] \in \tilde{X} \mid |u| < r\right\}$$

for some $r > 0$, and any neighborhood of q contains an open neighborhood of the type $\hat{U}_q = \left\{\left[(u', 2)\right] \in \tilde{X} \mid \left|u'\right| < r'\right\}$ for some $r' > 0$. We thus see that the intersection of any two neighborhoods of p and q is non-empty.

2. We prove next that $\tilde{X}$ is not compact. Indeed, for any $n \in \mathbb{Z} - \{0\}$, consider the open set

$$U_n = \left\{[(x, 1)] = [(x, 2)] \in \tilde{X} \mid n - 1 < x < n + 1\right\}$$

and consider also the open sets

$$U_0' = \left\{[(x, 1)] \in \tilde{X} \mid -1 < x < 1\right\}$$

and

$$U_0'' = \left\{[(x, 2)] \in \tilde{X} \mid -1 < x < 1\right\}.$$

The family $\mathcal{U} = \{U_n\}_{n \in \mathbb{Z}-\{0\}} \cup \left\{U_0', U_0''\right\}$ provides an open covering of $\tilde{X}$. But, for any $m \in \mathbb{Z}$, the point $[(m, y)] \in \tilde{X}$, for $y \in \{1, 2\}$, belongs to just one of the elements of the family $\mathcal{U}$. Hence there exists no finite sub-family of $\mathcal{U}$ covering $\tilde{X}$. This proves that $\tilde{X}$ is not compact.

3. We prove that $\tilde{X}$ is path-connected. Indeed, consider a point $[(x, y)] \in \tilde{X}$ with $x \in \mathbb{R}$ and $y \in \{1, 2\}$. Consider the path

$$\gamma\colon [0, 1] \to \tilde{X}, \quad \gamma(t) = [((1 - t)\, x + t,\ y)].$$

Note that $\gamma\colon [0, 1] \to \tilde{X}$ is continuous and that

$$\gamma(0) = [(x, y)]$$

while

$$\gamma(1) = [(1, 1)] = [(1, 2)].$$

Hence, any point $[(x, y)] \in \tilde{X}$ can be connected to the point $[(1, 1)] = [(1, 2)]$ by means of a continuous path in $\tilde{X}$.

7.16 Note that X_1 and X_2 are not compact, while X_3 and X_4 are. Hence $X_i \not\cong X_j$ for $i = 1, 2$ and $j = 3, 4$. Further, note that $X_1 \cong X_2$ by means of the stereographic projection

$$\varphi \colon X_1 \to X_2$$
$$(x, y) \mapsto \left(\frac{2x}{1+x^2+y^2}, \frac{2y}{1+x^2+y^2}, \frac{-1+x^2+y^2}{1+x^2+y^2} \right)$$
$$\varphi^{-1} \colon X_2 \to X_1$$
$$(x, y, z) \mapsto \left(\frac{x}{1-z}, \frac{y}{1-z} \right)$$

Finally, note that $X_3 \not\cong X_4$. Indeed, the property that a topological space remains connected after removing two points is a topological invariant. But if one removes two points from X_4, then the space obtained is connected while any two points one removes from X_3 results in a disconnected space.

7.17 Let S be an uncountable set and endow it with the cocountable topology. In this topology every countable subset is closed since its complement obviously has countable complement. In the other direction, if F is a closed subset of S, then its complement $S - F$ is open, and thus either $F = S$, or the complement of $S - F$, which is F, is countable. We thus identified the closed sets to be

$$\mathscr{F} = \{F \subseteq S \mid F \text{ is countable}\} \cup \{S\}.$$

Since a countable union of countable sets is countable, it follows that the collection $\mathscr{F}$ is stable under countable unions. To see that taking arbitrary unions may take one outside of the collection $\mathscr{F}$ it suffices to observe that a proper uncountable subset of S exists. Indeed, let $X \subset S$ be such a set. Since

$$X = \bigcup_{x \in X} \{x\}$$

the set X is certainly a union of closed sets. However, since X is uncountable and $X \neq S$, it follows that S is not closed. We leave it to the reader to exhibit proper uncountable subsets of S.

7.18 Since A is a real symmetric matrix, it is diagonalizable. It follows that, up to a homeomorphism induced by a change of coordinates of $\mathbb{R}^n$, we can suppose that A is a diagonal matrix with positive entries $\lambda_1 > 0, \ldots, \lambda_n > 0$,

$$A = \begin{pmatrix} \lambda_1 & & \\ & \ddots & \\ & & \lambda_n \end{pmatrix},$$

and so

$$W = \left\{ \begin{pmatrix} x_1 \\ \vdots \\ x_n \end{pmatrix} \in \mathbb{R}^n \mid \lambda_1 x_1^2 + \cdots + \lambda_n x_n^2 = 1 \right\}.$$

The set W is closed and bounded, and hence compact, in $\mathbb{R}^n$ with the Euclidean topology.

7.19 We prove that D is open in X. Take any point $p \in D$. Since D is locally compact, there exists a compact set $K \subseteq D$ in D and an open set U in X such that

$$p \in U \cap D \subseteq K.$$

Consider the set $U - K \subseteq X$, which is open in X since K is compact in the Hausdorff space X, and is thus closed. Since D is dense, if the open set $U - K$ were non-empty, it would intersect D, but this contradicts with

$$U \cap D \subseteq K.$$

Hence $U - K = \emptyset$, that is

$$p \in U \subseteq K \subseteq D.$$

Hence the point p is an interior point of D. As p was arbitrary, we conclude that all of the points of D are interior, and thus that D is open.

7.20 1. We have to show that $\mathscr{V}$ contains the empty set and the whole space X, and that it is closed under finite unions and under arbitrary intersections. Indeed, $\varnothing = V(1 \cdot \mathbb{Z}) \in \mathscr{V}$ and $X = V(0 \cdot \mathbb{Z}) \in \mathscr{V}$. Take $V(a_1 \cdot Z) \in \mathscr{V}$ and $V(a_2 \cdot \mathbb{Z}) \in \mathscr{V}$. Note that

$$V(a_1 \cdot \mathbb{Z}) \cup V(a_2 \cdot \mathbb{Z}) = V(\text{lcm}\{a, b\} \cdot \mathbb{Z}) \in \mathscr{V}.$$

Consider the family $\{V(a_n \cdot \mathbb{Z})\}_{n \in \mathbb{Z}} \subseteq \mathscr{V}$. Note that

$$\bigcap_{n \in \mathbb{Z}} V(a_n \cdot \mathbb{Z}) = V(\gcd\{a_n : n \in \mathbb{Z}\}) \in \mathscr{V}.$$

2. We prove that $\{0 \cdot \mathbb{Z}\} \subset X$ is dense in X, that is, $\overline{\{0 \cdot \mathbb{Z}\}} = X$. Indeed, the only closed set containing $\{0 \cdot \mathbb{Z}\}$ is $V(0 \cdot \mathbb{Z}) = X$. In particular, X is the smallest closed set containing $\{0 \cdot \mathbb{Z}\}$, which is precisely the closure.
3. We prove now that X is not Hausdorff. Indeed, note that for any $a \in \mathbb{Z}$ the closed set $V(a \cdot \mathbb{Z}) = \{p \cdot \mathbb{Z} \mid p \text{ is a prime dividing } a \text{ or } p = a = 0\}$ is a finite set. Hence, any two open sets in X intersect and hence X is not T_2.

7.6.3 Metric Spaces

7.21 Recall that in a metric space every compact set is also sequentially compact. Since $\{x_n\}_{n\in\mathbb{N}}$ does not admit any convergent subsequence, the space (X, d) cannot be sequentially compact, and consequently is not compact. Another approach to this problem is to use the fact that a metric space is compact if, and only if, it is complete and totally bounded. But clearly X is not totally bounded (though it may be bounded and it may be complete).

7.22 To show that ℓ_∞ is connected we actually show that it is path-connected and thus consider an arbitrary bounded real sequence $x = (x_n)_n \in \ell_\infty$, for which it suffices to construct a path to the 0 vector, namely the constantly 0 sequence. To that end, consider the function

$$\gamma : [0, 1] \to \ell_\infty, \quad \gamma(t) = (tx_n)_n .$$

Notice that the codomain is indeed ℓ_∞, as is immediately seen (in fact this is nothing but the closure of ℓ_∞ under scalar products). Since $\gamma(1) = x$ and $\gamma(0) = 0$, it remains to show that γ is continuous with respect to the relevant metrics. And indeed, for all $t, s \in [0, 1]$, it holds that

$$d(\gamma(t), \gamma(s)) = d\left((tx_n)_n, (sx_n)_n\right) = \sup_n |(t-s)\cdot x_n| = |t-s|\cdot \sup_n |x_n| = \|x\|_\infty \cdot d(s,t)$$

and thus γ is uniformly continuous (notice that we actually proved γ is Lipschitz), and thus continuous, as was needed.

To see that ℓ_∞ as not compact it suffices to construct a sequence of infinitely many vectors with all pair-wise distances equal to 1. This can easily be done in many ways. We mention here another approach. Notice that the set of all vectors of the form $x = (x_1, 0, 0, 0, \ldots)$, with $x_1 \in \mathbb{R}$ arbitrary, in other words the kernel of the shift mapping $f : \ell_\infty \to \ell_\infty$, is a closed subspace of ℓ_∞ which, with the induced metric, is isometric to $\mathbb{R}$ with the Euclidean metric. Since ℓ_∞ is Hausdorff (any metric space is Hausdorff), if ℓ_∞ were compact, then $\mathbb{R}$ would be compact as well (since any closed subspace of a compact Hausdorff space is compact). But $\mathbb{R}$ is of course not compact.

7.23 For all $n \geq 0$ let $X_n = \{x \in \mathbb{R} \mid f^{(n)} = 0\}$. Since f is infinitely differentiable, $f^{(n)}$ is continuous, and thus the set X_n, being the inverse image of the closed set $\{0\}$, is closed. Moreover, the condition on the function f states precisely that

$$\mathbb{R} = \bigcup_{n\geq 0} X_n$$

and thus we have expressed the non-empty and complete metric space $\mathbb{R}$ (with the Euclidean metric) as a countable union of closed sets. By Baire's Theorem at least one of the sets X_n contains a non-empty open set. Thus, there exists an $n \geq 0$ and an open set $U \neq \emptyset$ with $U \subseteq X_n$. But any open set in $\mathbb{R}$, by definition, contains an open interval. Thus, there exist $a < b$ with $(a, b) \subseteq X_n$. In other words, $f^{(n)}(x) = 0$ for all $x \in (a, b)$. Integrating n times in the range a to b reveals that f is a polynomial of degree at most n.

7.24 Recall that in any inner product space the parallelogram law

$$\|x + y\|^2 + \|x - y\|^2 = 2\|x\|^2 + 2\|y\|^2$$

holds for all vectors x and y. We prove that there exists no inner product on $C([0, 1], \mathbb{R})$ whose associated norm is $\|\cdot\|_\infty$ by showing the parallelogram law fails. Indeed, consider, for example, the continuous functions $x, y : [0, 1] \to \mathbb{R}$ given by

$$x(t) = 1 \quad \text{and} \quad y(t) = t.$$

One computes that

$$\begin{aligned} \|x + y\|_\infty^2 + \|x - y\|_\infty^2 &= \max_{t\in[0,1]} |x(t) + y(t)|^2 + \max_{t\in[0,1]} |x(t) - y(t)|^2 \\ &= \max_{t\in[0,1]} |1 + t|^2 + \max_{t\in[0,1]} |1 - t|^2 = 4 + 1 = 5 \end{aligned}$$

while

$$\begin{aligned} 2\|x\|_\infty^2 + 2\|y\|_\infty^2 &= 2 \max_{t\in[0,1]} |x(t)|^2 + 2 \max_{t\in[0,1]} |y(t)|^2 \\ &= 2 \max_{t\in[0,1]} |1|^2 + 2 \max_{t\in[0,1]} |t|^2 = 2 \cdot 1 + 2 \cdot 1 = 4. \end{aligned}$$

Therefore, the norm $\|\cdot\|$ does not satisfy the parallelogram law and is thus not induced by any inner product.

7.25 A metric on the empty set is a function $d : \emptyset \times \emptyset \to \mathbb{R}_+$ satisfying the metric axioms. The domain is the empty set, and thus there is a unique such function, which is vacuously satisfying the metric axioms. As for a metric on a singleton set $X = \{p\}$, that is a function $X \times X \to \mathbb{R}_+$, notice that the axioms require that $d(p, p) = 0$, and thus d is already determined. It is trivially seen that this d indeed determines a metric on X. Finally, if X contains at least two distinct points p an q, then noting first that there always exists a metric structure on d, for instance $d(x, y) = 1$ for all $x \neq y$ and $d(x, x) = 0$ for all x, the existence of infinitely many non-isometric metric structures follows by noting that if d is any metric function and $\alpha > 0$ an arbitrary positive real number, then defining

$$\alpha d : X \times X \to \mathbb{R}_+$$

by $(\alpha d)(x, y) = \alpha d(x, y)$ is also a metric on X. Clearly the spaces (X, d) and $(X, \alpha d)$ are isometric if, and only if, $\alpha = 1$ or if $d(x, y) = 0$ for all points. However, since $p \neq q$ we have that $d(p, q) \neq 0$. We thus obtain infinitely many non-isometric metrics on X.

7.26 Suppose that $x \in X$ satisfies $d(x, S) = 0$ and let U be an arbitrary open set with $x \in U$. It then holds that $B_\varepsilon(x) \subseteq U$ for some $\varepsilon > 0$. Since $d(x, S) = 0$ it follows that there is an element $s \in S$ with $d(x, s) < \varepsilon$. In particular then $s \in B_\varepsilon(x) \subseteq U$. We thus showed that every open set containing x intersects S, and thus $x \in \overline{S}$.

Conversely, if $x \notin \overline{S}$, then (noting that $\overline{S}$ is closed, and thus its complement is open) there exists an $\varepsilon > 0$ with $B_\varepsilon(x) \subseteq X - S$. But then $d(x, s) \geq \varepsilon$ for all $s \in S$, and thus $d(x, S) \geq \varepsilon$.

7.27 We establish the metric space axioms for $\tilde{d}$. Clearly, $\tilde{d}(x, y) \geq 0$ for all $x, y \in X$. If $\tilde{d}(x, y) = 0$, then $d(x, y) = 0$ and thus $x = y$. For all $x, y \in X$ if $d(x, y) < 1$, then

$$\tilde{d}(x, y) = d(x, y) = d(y, x) = \tilde{d}(y, x)$$

while if $d(x, y) \geq 1$, then $d(y, x) \geq 1$ too, and thus $\tilde{d}(x, y) = 1 = \tilde{d}(y, x)$. Finally, the verification of the triangle inequality follows similarly by case splitting.

To show that d and $\tilde{d}$ induce the same topology, it suffices to notice that an open ball $B_\varepsilon(x)$ of radius $\varepsilon < 1$ is the same set when computed in (X, d) as it is when computing in $(X, \tilde{d})$.

7.28 We prove that there exists no isometry from $\mathbb{R}^2$ to $\mathbb{R}$ when endowed with the Euclidean metrics $d_{\mathbb{R}^2}$ and $d_{\mathbb{R}}$ respectively. Indeed, suppose that such an isometry $f: \mathbb{R}^2 \to \mathbb{R}$ exists. Take $P, Q, R \in \mathbb{R}^2$ such that

$$d_{\mathbb{R}^2}(P, Q) = d_{\mathbb{R}^2}(P, R) = d_{\mathbb{R}^2}(Q, R) = 1.$$

Denote by $p = f(P), q = f(Q)$, and $r = f(R)$ the images in $\mathbb{R}$ of the three chosen points. Since f is an isometry, it follows that

$$|p - q| = d_{\mathbb{R}}(p, q) = d_{\mathbb{R}}(f(P), f(Q)) = d_{\mathbb{R}^2}(P, Q) = 1,$$

and analogously that $|p - r| = 1$ and $|q - r| = 1$. However, elementary algebra reveals this to be an impossibility.

7.29 We show that d_s satisfies the metric axioms, remembering that each d_i does. Firstly, for all $x \in X$

$$d_s(x, x) = \sup_{i \in I} d_i(x, x) = \sup_{i \in I}\{0\} = 0.$$

Next, for all $x, y \in X$

$$d_s(x, y) = \sup_{i \in I}\{d_i(x, y)\} = \sup_{i \in I}\{d_i(y, x)\} = d_s(y, x).$$

Finally, for all $x, y, z \in X$, we need to show that

$$\sup_{i\in I}\{d_i(x,z)\} \le \sup_{i\in I}\{d_i(x,y)\} + \sup_{i\in I}\{d_i(y,z)\}$$

for which it suffices to show, for each $i \in I$, that

$$d_i(x,z) \le \sup_{i\in I}\{d_i(x,y)\} + \sup_{i\in I}\{d_i(y,z)\}$$

And indeed, given $i \in I$, using the triangle inequality for d_i

$$d_i(x,z) \le d_i(x,y) + d_i(y,z) \le \sup_{i\in I}\{d_i(x,y)\} + \sup_{i\in I}\{d_i(y,z)\},$$

completing the proof.

For an example showing that the infimum of metric functions need not be a metric function, consider a set $X = \{a, b, c\}$ with three distinct points. On it consider the following assignments of distances

$$\begin{array}{ll} d_1(a,b) = 1 & d_2(a,b) = 2 \\ d_1(b,c) = 2 & d_2(b,c) = 1 \\ d_1(a,c) = 3 & d_2(a,c) = 3 \end{array}$$

which are easily seen to endow X with two structures of metric space. However, when computing the minima of the distances one obtains the function e with $e(a,b) = e(b,c) = 1$, and $e(a,c) = 3$. Therefore $e(a,c) > e(a,b) + e(b,c)$ and so the triangle inequality fails.

7.30
1. No, not continuity nor uniform continuity suffices. For instance, let $X = \mathbb{R}$ with the Euclidean topology and consider $f : \mathbb{R} \to \mathbb{R}$ with $f(x) = \arctan(x)$, which is uniformly continuous, and thus also continuous. Then $\mathbb{R}$ is complete but the set $f(x) = (-\pi/2, \pi/2)$ is not complete.
2. Uniform continuity is a sufficient condition. Indeed, to show that $f(x)$ is totally bounded, let $\varepsilon > 0$ be given and we will find a covering of $f(x)$ by sets whose diameters do not exceed ε. Let $\delta > 0$ correspond to the given ε, that is $\mathrm{diam}(f(S)) < \varepsilon$ whenever $S \subseteq X$ satisfies $\mathrm{diam}(S) < \delta$. Since X is totally bounded, one can cover X by sets $S_1, \ldots, S_m$ whose diameters do not exceed δ. Each of the images $f(S_1), \ldots, f(S_m)$ has diameter not exceeding ε, and they cover $f(x)$, as needed. Finally, we note that continuity alone does not suffice. For instance, consider the function $\tan : (-\pi/2, \pi/2) \to \mathbb{R}$, with $\mathbb{R}$ endowed with the Euclidean metric and the interval with the induced metric. The interval is totally bounded but its image, $\mathbb{R}$, is not even bounded.
3. Continuity, and thus also uniform continuity, suffices. Indeed, in a metric space a set is compact if, and only if, it is complete and totally bounded. Moreover, for a continuous function f, the image $f(C)$ is compact whenever C is.

7.6.4 Normed Spaces and Banach Spaces

7.31 First we show that the metric $d(x, y) = \|x - y\|$ induced by the norm of a normed space V is translation invariant and scale homogeneous. Indeed,

$$d(x + z, y + z) = \|(x + z) - (y + z)\| = \|x - z\| = d(x, z)$$

holds for all $x, y, z \in V$ and for any scalar α

$$d(\alpha x, \alpha y) = \|\alpha x - \alpha y\| = \|\alpha(x - y)\| = |\alpha|\|x - y\| = |\alpha|d(x, y).$$

In the other direction, suppose that V is endowed with a translation invariant and scale homogeneous metric d. We then define, for all $x \in V$

$$\|x\| = d(0, x)$$

and proceed to show that the axioms for a normed space are satisfied. It is clear that $\|x\| \geq 0$ for all $x \in V$ and that if $\|x\| = 0$, then $d(0, x) = 0$, and thus $x = 0$. For any scalar α and $x \in V$ one has

$$\|\alpha x\| = d(0, \alpha x) = d(\alpha \cdot 0, \alpha x) = |\alpha|d(0, x) = |\alpha|\|x\|.$$

Finally, for all $x, y \in V$ we have, by translation invariance, that $d(x, x+y) = d(0, y)$. It then follows that

$$\|x + y\| = d(0, x + y) \leq d(0, x) + d(x, x + y) = d(0, x) + d(0, y) = \|x\| + \|y\|,$$

as required.

7.32 1. We establish the norm axioms for the given proposed norm function. Clearly $\|f\| \geq 0$ since the integral of a non-negative function is non-negative. Moreover, suppose that $\|f\| = 0$ for a continuous function $f : [a, b] \to \mathbb{R}$. If $f \neq 0$, then $f(x_0) \neq 0$ for some $x_0 \in [a, b]$. Since $|f(x)|$ is continuous and $|f(x_0)| > 0$ it follows that there is an interval $(c, d) \subseteq [a, b]$ upon which $|f(x)|$ attains values greater than $|f(x_0)|/2 > 0$. It follows that

$$\int_a^b |f(x)|dx \geq \int_c^d |f(x)|dx \geq (d - c)|f(x_0)|/2 > 0$$

which is a contradiction. We conclude thus that $\|f\| = 0$ implies $f = 0$. Next, for all $\lambda \in \mathbb{R}$

$$\|\lambda \cdot f\| = \int_a^b |\lambda \cdot f(x)|\, dx = \lambda \cdot \int_a^b |f(x)|\, dx = \lambda \cdot \|f\|\,.$$

Finally, for $f\colon [a, b] \to \mathbb{R}$ and $g\colon [a, b] \to \mathbb{R}$, it holds that

$$\|f + g\| = \int_a^b |f(x) + g(x)|\, dx \le \int_a^b |f(x)|\, dx + \int_a^b |g(x)|\, dx = \|f\| + \|g\|\,,$$

establishing the triangle inequality.

2. We prove that the metric space (X, d) with $d(f, g) = \|f - g\|$ is not complete. For simplicity, assume that $[a, b] = [-1, 1]$. We provide an example of a Cauchy sequence in X having no limit point in X. For all $n \in \mathbb{N}$, consider the function

$$f_n\colon [0, 1] \to \mathbb{R}, \quad f_n(x) = \begin{cases} -1 & \text{for } -1 \le x < -\frac{1}{n} \\ nx & \text{for } -\frac{1}{n} \le x < \frac{1}{n} \\ 1 & \text{for } \frac{1}{n} \le x \le 1. \end{cases}$$

Note that $f_n(x) \in X$ for all $n \in \mathbb{N}$. Furthermore, for $n, m \in \mathbb{N}$ with $n < m$ one computes

$$\begin{aligned} \|f_n - f_m\| &= \int_{-1}^{1} |f_n(x) - f_m(x)|\, dx \\ &= 2 \left(\int_0^{\frac{1}{m}} (m - n)\, x\, dx + \int_{\frac{1}{m}}^{\frac{1}{n}} (1 - nx)\, dx \right) \\ &\le 2 \int_0^{\frac{1}{n}} (1 - nx)\, dx = \frac{1}{n}. \end{aligned}$$

It thus follows that the sequence $(f_n)_{n\in\mathbb{N}}$ is a Cauchy sequence in (X, d). We prove now that $(f_n)_{n\in\mathbb{N}}$ has no limit point in X. Indeed, we claim that, if such a limit point $f \in X$ were to exist, then it would be that

$$f(x) = -1 \text{ for all } x \in [-1, 0) \quad \text{and} \quad f(x) = 1 \text{ for all } x \in (0, 1],$$

which is absurd since such a function f is not continuous on $[-1, 1]$. To prove the claim, suppose that there were a point $x \in [-1, 0)$ such that $|f(x) - (-1)| = \delta$ for some $\delta > 0$. For simplicity, we assume $x \neq -1$: otherwise, by continuity, if $x = -1$, then there exists a point $x' > -1$ such that $\left|f(x') - (-1)\right| = \delta' > 0$.

Then, by continuity, there would be a neighbourhood $(x - \varepsilon, x + \varepsilon) \subseteq [-1, 0)$, where $\varepsilon > 0$, for x in $[-1, 0)$ such that $|f(y) - (-1)| > \delta/2$ for all $y \in (x - \varepsilon, x + \varepsilon)$. Choose $N > -1/(x + \varepsilon)$, so $x + \varepsilon < -1/N$. Hence, for all $n \geq N$, it holds that

$$\|f_n - f\| = \int_{-1}^{1} |f_n(x) - f(x)|\, dx \geq \int_{x-\varepsilon}^{x+\varepsilon} |f_n(x) - f(x)|\, dx \geq 2\varepsilon\delta > 0,$$

which is not possible.

7.33 We first prove that d is a distance on $\mathbb{R}$. Firstly, it is clear that for all $x, y \in \mathbb{R}$

$$0 \leq d(x, y) = d(y, x).$$

Furthermore, if $d(x, y) = 0$ then $|x - y| = 0$, and thus $x = y$. For the triangle inequality, consider the function $f : [0, \infty)] \to [0, \infty]$ given by

$$f(t) = \frac{t}{1 + t}$$

and notice that the proposed distance function is

$$d(x, y) = f(d_E(x, y))$$

where $d_E(x, y) = |x - y|$ is the Euclidean metric on $\mathbb{R}$. Using elementary analysis it is seen that f is monotonically increasing and that it is concave and thus it is also subadditive, i.e.,

$$f(s + t) \leq f(s) + f(t).$$

We may now argue as follows. For all $x, y, z \in \mathbb{R}$

$$d(x, z) = f(d_E(x, z)) \leq f(d_E(x, y) + d_E(y, z))$$

where we used the monotonicity of f. The subadditivity of f now implies that

$$f(d_E(x, y) + d_E(y, z)) \leq f(d_E(x, y)) + f(d_E(y, z)) = d(x, y) + d(y, z)$$

establishing the triangle inequality, and thus that $(\mathbb{R}, d)$ is a metric space.

We now show that d is not induced by any norm. Indeed, if there were such a norm $x \mapsto \|x\|$ for which $d(x, y) = \|x - y\|$, then, for all $x, y \in \mathbb{R}$ and all $\lambda \in \mathbb{R}$, it would hold that

$$d(\lambda x, \lambda y) = \|\lambda x - \lambda y\| = |\lambda| \cdot \|x - y\| = |\lambda| \cdot d(x, y).$$

However, by using the explicit expression of d, we get

$$\frac{|\lambda x - \lambda y|}{1 + |\lambda x - \lambda y|} = \frac{|x - y|}{1 + |x - y|}$$

and taking, e.g., $x = 1$, $y = 0$, and $\lambda = 2$, the left-hand-side is $2/3$ while the right-hand-side is $1/2$, yielding an absurdity.

7.34 To show that the topology induced by $\|\cdot\|_2$ is finer than the topology induced by $\|\cdot\|_1$ it suffices to show that the identity function

$$\mathrm{id} : (X, \|\cdot\|_2) \to (X, \|\cdot\|_1)$$

is continuous. And indeed, given $\varepsilon > 0$ let $\delta = \varepsilon/K$. Then, for all $x_1, x_2 \in X$ such that $\|x_1 - x_2\|_2 < \delta$, one has $\|\mathrm{id}(x_1) - \mathrm{id}(x_2)\|_1 = \|x_1 - x_2\|_1 \leq K \cdot \|x_1 - x_2\|_2 < K\delta = \varepsilon$.

7.35 1. Given $x, y \in V$ with $\|x\| = \|y\| = 1$, for every $0 \leq \alpha \leq 1$ the triangle inequality gives

$$\|\alpha x + (1 - \alpha)y\| \leq \|\alpha x\| + \|(1 - \alpha)y\| = \alpha\|x\| + (1 - \alpha)\|y\| = 1.$$

2. Suppose now that $\|x\|^2 = \langle x, x\rangle$ for an inner product $\langle\cdot, \cdot\rangle$ on V, and let's assume the ground field is $\mathbb{R}$ (the case $K = \mathbb{C}$ is left as an extra exercise). If $x \neq y$ satisfy $\|x\| = \|y\| = 1$, then for all $0 < \alpha < 1$

$$\|\alpha x + (1 - \alpha)y\|^2 = \alpha^2 + (1 - \alpha)^2 + 2\alpha(1 - \alpha)\langle x, y\rangle.$$

By the Cauchy-Schwarz Inequality we have that

$$|\langle x, y\rangle| \leq \|x\|\|y\| = 1.$$

Since $x \neq y$ and $\|x\| = \|y\|$, it follows that x is not a scalar multiple of y, and thus the Cauchy-Schwarz Inequality holds strictly. Thus $\langle x, y\rangle < 1$ and it follows that

$$\|\alpha x + (1 - \alpha)y\|^2 < 1 + 2\alpha^2 - 2\alpha + 2\alpha - 2\alpha^2 = 1$$

showing that $\|\alpha x + (1 - \alpha)y\| < 1$ and thus the only points of intersection of $L(x, y)$ with the unit sphere occur when $\alpha = 0$ or $\alpha = 1$ at the points y and x, respectively.

3. Assume that $x \neq y$ satisfy $\|x\| = \|y\| = 1$ in a strictly convex space. By the triangle inequality $\|x + y\| \leq \|x\| + \|y\| \leq 2$ so we just need to show that $\|x + y\| = 2$ is impossible. Indeed, if $\|x + y\| = 2$, then

$$\|(1/2)x + (1/2)y\| = (1/2)\|x + y\| = 1$$

contradicting strict convexity.

4. $\mathbb{R}^2$ with the ℓ_2 norm is strictly convex since the ℓ_2 norm is induced by an inner product (the standard inner product). That $\mathbb{R}^2$ with the ℓ_1 norm is not strictly convex is seen by choosing $x = (1, 0)$ and $y = (0, 1)$. Then $\|x\|_1 = \|y\|_1 = 1$ but $\|x + y\| = 2$. Similarly, for $x = (1, 1)$ and $y = (1, -1)$ it holds that $\|x\|_\infty = \|y\|_\infty = 1$, but $\|x + y\|_\infty = 2$, showing that the ℓ_∞ norm is not strictly convex either.

7.36 To show that the kernel of A is a closed linear subspace of V recall that the kernel, by definition, is the set $A^{-1}(\{0\})$. The fact that the kernel is a linear subspace of the domain was established in the main text and in any case is easy to re-establish if needed. To show that the kernel is closed recall that the inverse image under a continuous function of a closed set is closed. Since a bounded linear operator is continuous, it suffices to show that $\{0\}$ is a closed set in W. Indeed, the metric induced on W by the norm is a metrizable topology and thus is Hausdorff, and thus every singleton set is closed.

7.37 The verification that A is a linear operator is trivial, and thus we omit it. Next we observe that for all $x \in \ell_2$ with $x \neq 0$

$$\|Ax\|^2 = \sum_{k=1}^{\infty} |(1 - \frac{1}{n})x_n|^2 < \sum_{k=1}^{\infty} |x_k|^2 = \|x\|^2$$

and thus

$$\|Ax\| < \|x\|$$

holds for all $x \neq 0$. It follows that $\|A\| \leq 1$ and that if $\|A\| = 1$, then the norm is never attained. To see that the operator norm of A is indeed precisely 1, consider

$$e_n = (0, 0, \ldots, 0, 1, 0, 0, \ldots)$$

with 1 in the n-th position. It holds that

$$\|Ae_n\| = 1 - \frac{1}{n}$$

and since $\|e_n\| = 1$ it follows that $\|A\| \geq \|Ae_n\|$, for all $n \geq 1$, and so we conclude that indeed $\|A\| = 1$.

7.38 The composition of linear operators is always a linear operator, so all we need to do is show the composition BA is bounded, which will be done by showing the requested upper bound of the operator norm. Indeed, for all $x \in U$

$$\|BAx\| = \|B(Ax)\| \leq \|B\|\|Ax\| \leq \|B\|\|A\|\|x\|$$

and the claim follows.

7.39 In an exercise the reader was requested to prove that for bounded linear operators $A, B \in \mathbf{B}(\mathscr{B})$, if A is invertible and $\|A - B\| < \|A\|$, then B is invertible. With this result in mind one just needs to correctly interpret the meaning of the claim that G is open. Let $A \in G$ be given. To show that G is open it suffices to show that G contains an open ball with positive radius and centre A. In other words, we need to find $\varepsilon > 0$ such that every $B \in \mathbf{B}(\mathscr{B})$ with $\|A - B\| < \varepsilon$ is invertible. By the above result, one may take $\varepsilon = \|A\|$, which is positive (since A is invertible, so clearly $A \neq 0$).

7.40 Notice that the domain of ST is the set $\{x \in \mathscr{B}_1 \mid Tx \in \mathscr{D}(S)\}$. Given a sequence $\{x_n\}$ in the domain of ST, suppose that $x_n \to x_0$ and that $STx_n \to z$. We need to show that x_0 is in the domain of ST and that $STx_0 = z$. Indeed, since T is continuous, it follows that

$$Tx_n \to Tx_0.$$

Since Tx_n is in the domain of S and since $S(Tx_n) \to z$, it follows from the fact that S is closed that Tx_0 is in the domain of S and that $STx_0 = z$. Thus, as was required, we see that x_0 is in the domain of S and that $STx_0 = z$, showing ST is closed.

7.6.5 Topological Groups

7.41 Firstly, note that $\mathbb{R}^n$ with the Euclidean topology is a Hausdorff topological space since the topology is merizable. Both the group sum

$$+\colon \mathbb{R}^n \times \mathbb{R}^n \to \mathbb{R}^n, \quad ((a_1, \dots, a_n), (b_1, \dots, b_n)) \overset{+}{\mapsto} (a_1 + b_1, \dots, a_n + b_n)$$

and the group inverse

$$-\colon \mathbb{R}^n \to \mathbb{R}^n, \quad (a_1, \dots, a_n) \mapsto (-a_1, \dots, -a_n)$$

are continuous with respect to the Euclidean topology on $\mathbb{R}^n$ and to the product topology on $\mathbb{R}^n \times \mathbb{R}^n$. Hence, the additive group $\mathbb{R}^n$ with the Euclidean topology is a topological group. In fact, the map

$$\mathbb{R} \times \mathbb{R}^n \to \mathbb{R}^n, \quad (\lambda, (a_1, \dots, a_n)) \overset{\cdot}{\mapsto} (\lambda \cdot a_1, \dots, \lambda \cdot a_n)$$

is also continuous with respect to the Euclidean topology of $\mathbb{R}^n$ and the product topology of $\mathbb{R} \times \mathbb{R}^n$. That is to say, the $\mathbb{R}$-vector space $\mathbb{R}^n$ is actually a topological vector space.

7.42 Note that $\mathrm{SL}_n(\mathbb{R}) = \det^{-1}(\{1\})$, and $\det\colon \mathrm{GL}_n(\mathbb{R}) \to \mathbb{R}$ is continuous. Hence $\mathrm{SL}_n(\mathbb{R})$ is a closed subgroup of the topological group $\mathrm{GL}_n(\mathbb{R})$. In particular, $\mathrm{SL}_n(\mathbb{R})$ is a topological group.

7.43 Note that $\mathrm{O}(n) = \varphi^{-1}(\{\mathrm{id}_n\})$, where $\varphi : \mathrm{GL}_n(\mathbb{R}) \to \mathrm{GL}_n(\mathbb{R})$ is given by

$$\varphi(A) = A^{\mathrm{t}}A,$$

and is a continuous map. Hence $\mathrm{O}(n)$ is a closed subgroup of the topological group $\mathrm{GL}_n(\mathbb{R})$. In particular, $\mathrm{O}(n)$ is a topological group.

7.44 Note that $\mathrm{SO}(n) = \det^{-1}(\{1\})$, where $\det\colon \mathrm{O}(n) \to \{1, -1\}$ is continuous. Hence $\mathrm{SO}(n)$ is a closed subgroup of the topological group $\mathrm{O}(n)$ and is thus a topological group.

7.45 Note that $\mathrm{Sp}_{2n}(\mathbb{R}) = \varphi^{-1}(\{J_0\})$, where $\varphi\colon \mathrm{GL}_{2n}(\mathbb{R}) \to \mathrm{GL}_{2n}(\mathbb{R})$, given by

$$\varphi(A) = A^{\mathrm{t}} \cdot J_0 \cdot A,$$

is a continuous map. Hence $\mathrm{Sp}_{2n}(\mathbb{R})$ is a closed subgroup of the topological group $\mathrm{GL}_{2n}(\mathbb{R})$, and thus is itself a topological group.

7.46 For a given $B \in \mathrm{SO}(n)$, we construct a path

$$\gamma\colon [0,1] \to \mathrm{SO}(n) \quad \text{such that} \quad \gamma(0) = \mathrm{id}_n \text{ and } \gamma(1) = B.$$

The matrix B being orthogonal, there exists an orthogonal matrix M such that

$$B = M^{\mathrm{t}}KM$$

where

$$K = \begin{pmatrix} \mathrm{id}_p & & \\ & -\mathrm{id}_q & \\ & & S \end{pmatrix},$$

with $q = 2r$ and, for $k = n - (p+q)$,

$$S = \begin{pmatrix} \cos\alpha_1 & -\sin\alpha_1 & & & \\ \sin\alpha_1 & \cos\alpha_1 & & & \\ & & \ddots & & \\ & & & \cos\alpha_k & -\sin\alpha_k \\ & & & \sin\alpha_k & \cos\alpha_k \end{pmatrix}$$

for some $\alpha_1, \ldots, \alpha_k \in (0, \pi) \cup (\pi, 2\pi)$. For $t \in [0, 1]$, define

$$K_t = \left(\begin{array}{c|c|c} \mathrm{id}_p & & \\ \hline & R_t & \\ \hline & & S_t \end{array}\right),$$

where

$$R_t = \left(\begin{array}{c|c|c} \begin{matrix} \cos(t\pi) & -\sin(t\pi) \\ \sin(t\pi) & \cos(t\pi) \end{matrix} & & \\ \hline & \ddots & \\ \hline & & \begin{matrix} \cos(t\pi) & -\sin(t\pi) \\ \sin(t\pi) & \cos(t\pi) \end{matrix} \end{array}\right).$$

and

$$S_t = \left(\begin{array}{c|c|c} \begin{matrix} \cos(t\alpha_1) & -\sin(t\alpha_1) \\ \sin(t\alpha_1) & \cos(t\alpha_1) \end{matrix} & & \\ \hline & \ddots & \\ \hline & & \begin{matrix} \cos(t\alpha_k) & -\sin(t\alpha_k) \\ \sin(t\alpha_k) & \cos(t\alpha_k) \end{matrix} \end{array}\right).$$

Hence define, for all $t \in [0, 1]$,

$$\gamma(t) = M^{\mathrm{t}} K_t M \in \mathrm{SO}(n).$$

Since $\gamma : [0, 1] \to \mathrm{SO}(n)$ connects id_n to B, we get that $\mathrm{SO}(n)$ is path-connected and hence also connected.

7.47 Note that $\mathrm{O}(n)$ is the disjoint union of $\mathrm{O}^+(n)$ and $\mathrm{O}^-(n)$, where

$$\mathrm{O}^+(n) = \{A \in \mathrm{O}(n) \mid \det(A) > 0\}$$

and

$$\mathrm{O}^-(n) = \{A \in \mathrm{O}(n) \mid \det(A) < 0\}.$$

Furthermore, $\mathrm{O}^+(n)$ and $\mathrm{O}^-(n)$ are homeomorphic. In fact, the map

$$\mathrm{Or}^+(n) \to \mathrm{Or}^-(n), \quad \left(\begin{array}{c|ccc} x_{11} & x_{12} & \cdots & x_{1n} \\ \vdots & \vdots & \ddots & \vdots \\ x_{n1} & x_{n2} & \cdots & x_{nn} \end{array}\right) \mapsto \left(\begin{array}{c|ccc} -x_{11} & x_{12} & \cdots & x_{1n} \\ \vdots & \vdots & \ddots & \vdots \\ -x_{n1} & x_{n2} & \cdots & x_{nn} \end{array}\right)$$

provides a homeomorphism. Therefore, it suffices to show that $\mathrm{O}^+(n)$ is path-connected. But $\mathrm{O}^+(n) = \mathrm{SO}(n)$, and the claim follows since it was already established that $\mathrm{SO}(n)$ is path-connected.

7.48 For a given $B \in \mathrm{GL}_n^+(\mathbb{R})$, we construct a path

$$\gamma\colon [0,1] \to \mathrm{GL}_n^+(\mathbb{R}) \quad \text{such that} \quad \gamma(0) = \mathrm{id}_n \text{ and } \gamma(1) = B.$$

By the polar decomposition theorem there are a symmetric positive-definite matrix L and an orthogonal matrix P such that $B = LP$. Since $\det(L) > 0$ and P is orthogonal, it follows that $\det(P) = 1$, and hence $P \in \mathrm{O}^+(n) = \mathrm{SO}(n)$. The matrix L being symmetric, there exists an orthogonal matrix C such that

$$L = C^{\mathrm{t}} \Lambda C \quad \text{where} \quad \Lambda = \begin{pmatrix} \lambda_1 & & \\ & \ddots & \\ & & \lambda_n \end{pmatrix}.$$

For all $t \in [0,1]$ define

$$\Lambda_t = \begin{pmatrix} t\lambda_1 + (1-t) & & \\ & \ddots & \\ & & t\lambda_n + (1-t) \end{pmatrix}.$$

Since the topological group $\mathrm{SO}(n)$ is path-connected, there exists a path

$$\mu\colon [0,1] \to \mathrm{SO}(n) \quad \text{such that} \quad \mu(0) = \mathrm{id}_n \text{ and } \mu(1) = P.$$

Hence define, for all $t \in [0,1]$,

$$\gamma(t) = C^{\mathrm{t}} \Lambda_t C \mu(t).$$

Since $\gamma\colon [0,1] \to \mathrm{GL}_n^+(\mathbb{R})$ connects id_n to B, we get that $\mathrm{GL}_n^+(\mathbb{R})$ is path-connected and hence also connected.

7.49 Note that since the function $\det\colon \mathrm{M}_n(\mathbb{R}) \to \mathbb{R}$ is continuous and surjective, the sets

$$\mathrm{GL}_n^+(\mathbb{R}) = \det^{-1}((0,+\infty))$$

and

$$\mathrm{GL}_n^-(\mathbb{R}) = \det^{-1}((-\infty,0))$$

are non-empty open sets and

$$\mathrm{GL}_n^+(\mathbb{R}) \cap \mathrm{GL}_n^-(\mathbb{R}) = \emptyset$$

while

$$\mathrm{GL}_n(\mathbb{R}) = \mathrm{GL}_n^+(\mathbb{R}) \cup \mathrm{GL}_n^-(\mathbb{R}).$$

Note also that $\mathrm{GL}_n^+(\mathbb{R})$ and $\mathrm{GL}_n^-(\mathbb{R})$ are homeomorphic, in fact, the map

$$\mathrm{GL}_n^+(\mathbb{R}) \to \mathrm{GL}_n^-(\mathbb{R}), \quad \left(\begin{array}{c|ccc} x_{11} & x_{12} & \cdots & x_{1n} \\ \vdots & \vdots & \ddots & \vdots \\ x_{n1} & x_{n2} & \cdots & x_{nn} \end{array}\right) \mapsto \left(\begin{array}{c|ccc} -x_{11} & x_{12} & \cdots & x_{1n} \\ \vdots & \vdots & \ddots & \vdots \\ -x_{n1} & x_{n2} & \cdots & x_{nn} \end{array}\right)$$

provides a homeomorphism. The claim now follows since it was already proved that $\mathrm{GL}_n^+(\mathbb{R})$ is path-connected, and hence $\mathrm{GL}_n^-(\mathbb{R})$ is path-connected as well.

7.50 For a given $B \in \mathrm{SL}_n(\mathbb{R})$ we construct a path

$$\gamma\colon [0,1] \to \mathrm{SL}_n(\mathbb{R}) \quad \text{such that} \quad \gamma(0) = \mathrm{id}_n \text{ and } \gamma(1) = B.$$

By the polar decomposition theorem there exist a symmetric positive-definite matrix L and an orthogonal matrix P such that $B = LP$. In particular, since $\det(B) = 1$, it holds that $\det(P) = 1$ and so $\det(L) = 1$. The matrix L being symmetric, there exists an orthogonal matrix C such that

$$L = C^{\mathrm{t}} \Lambda C$$

and since L is positive-definite with determinant equal to 1, the diagonal matrix Λ has the form

$$\Lambda = \begin{pmatrix} \exp(\mu_1) & & \\ & \ddots & \\ & & \exp(\mu_n) \end{pmatrix},$$

for some $\mu_1, \ldots, \mu_n \in \mathbb{R}$ such that $\sum_{j=1}^n \mu_j = 0$. The matrix P being orthogonal, there exists an orthogonal matrix M such that

$$P = M^{\mathrm{t}} K M$$

where

$$K = \left(\begin{array}{c|c|c} \mathrm{id}_p & & \\ \hline & -\mathrm{id}_q & \\ \hline & & S \end{array}\right),$$

with $q = 2r$ and, for $k = n - (p+q)$,

$$S = \left(\begin{array}{cc|c|cc} \cos\alpha_1 & -\sin\alpha_1 & & & \\ \sin\alpha_1 & \cos\alpha_1 & & & \\ \hline & & \ddots & & \\ \hline & & & \cos\alpha_k & -\sin\alpha_k \\ & & & \sin\alpha_k & \cos\alpha_k \end{array}\right)$$

for some $\alpha_1, \ldots, \alpha_k \in (0, \pi) \cup (\pi, 2\pi)$. For all $t \in [0, 1]$ define

$$\Lambda_t = \begin{pmatrix} \exp(t\mu_1) & & \\ & \ddots & \\ & & \exp(t\mu_n) \end{pmatrix},$$

and define

$$K_t = \left(\begin{array}{c|c|c} \mathrm{id}_p & & \\ \hline & R_t & \\ \hline & & S_t \end{array}\right),$$

where

$$R_t = \left(\begin{array}{cc|c|cc} \cos(t\pi) & -\sin(t\pi) & & & \\ \sin(t\pi) & \cos(t\pi) & & & \\ \hline & & \ddots & & \\ \hline & & & \cos(t\pi) & -\sin(t\pi) \\ & & & \sin(t\pi) & \cos(t\pi) \end{array}\right).$$

and

$$S_t = \left(\begin{array}{cc|c|cc} \cos(t\alpha_1) & -\sin(t\alpha_1) & & & \\ \sin(t\alpha_1) & \cos(t\alpha_1) & & & \\ \hline & & \ddots & & \\ \hline & & & \cos(t\alpha_k) & -\sin(t\alpha_k) \\ & & & \sin(t\alpha_k) & \cos(t\alpha_k) \end{array}\right).$$

Hence, define for all $t \in [0, 1]$

$$\gamma(t) = C^{\mathrm{t}}\Lambda_t C M^{\mathrm{t}} K_t M \in \mathrm{SL}_n(\mathbb{R}).$$

Since $\gamma\colon [0, 1] \to \mathrm{SL}_n(\mathbb{R})$ connects id_n to B, we get that $\mathrm{SL}_n(\mathbb{R})$ is path-connected and hence connected.

$$S = \left(\begin{array}{c|c|c} & & \begin{matrix} \cos \alpha_1 & -\sin \alpha_1 \\ \sin \alpha_1 & \cos \alpha_1 \end{matrix} \\ \hline & \ddots & \\ \hline \begin{matrix} \cos \alpha_k & -\sin \alpha_k \\ \sin \alpha_k & \cos \alpha_k \end{matrix} & & \end{array}\right)$$

[illegible] For all $t \in [0, 1]$ define

$$\begin{pmatrix} \exp(t\lambda_1) & & \\ & \ddots & \\ & & \exp(t\lambda_l) \end{pmatrix}$$

and define

[illegible]

where

[illegible]

and

[illegible]

We define [illegible]

$$[illegible] M \in SL_n(\mathbb{R})$$

[illegible] $SL_n(\mathbb{R})$ is path-connected and hence connected.

Index

© Springer International Publishing Switzerland 2015
C. Alabiso and I. Weiss, *A Primer on Hilbert Space Theory*,
UNITEXT for Physics, DOI 10.1007/978-3-319-03713-4

K

L

M

N

Printed in the United States
By Bookmasters